Quantum physics: illusion or realit

Quantum physics: illusion or reality?

ALASTAIR I. M. RAE
Department of Physics
University of Birmingham

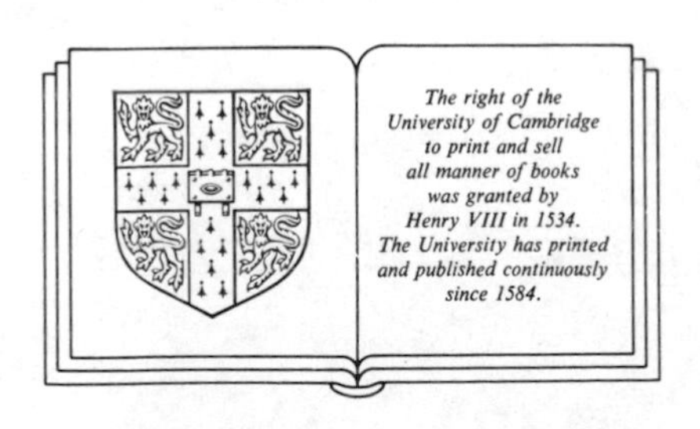

CAMBRIDGE UNIVERSITY PRESS
Cambridge
London New York New Rochelle
Melbourne Sydney

Published by the Press Syndicate of the University of Cambridge
The Pitt Building, Trumpington Street, Cambridge CB2 1RP
32 East 57th Street, New York, NY 10022, USA
10 Stamford Road, Oakleigh, Melbourne 3166, Australia

First published 1986
Reprinted 1986

Printed in Great Britain at the University Press, Cambridge

British Library cataloguing in publication data

Rae, Alastair I. M.
Quantum physics: illusion or reality.
1. Quantum theory
I. Title
530.1′2 QC174.12

Library of Congress cataloging in publication data

Rae, Alastair I. M.
Quantum physics: illusion or reality?
Bibliography
Includes index.
1. Quantum theory. 2. Physics – Philosophy. I. Title.
QC174.12.R335 1985 530.1′2 85–13256

ISBN 0 521 26023 X hard covers
ISBN 0 521 27802 3 paperback

PN

To Ann

I like relativity and quantum theories
Because I don't understand them
And they make me feel as if space shifted
About like a swan that can't settle
Refusing to sit still and be measured
And as if the atom were an impulsive thing
Always changing its mind.

D. H. Lawrence

Time present and time past
Are both perhaps present in time future
And time future contained in time past.

T. S. Eliot

Do you think the things people make fools of themselves about are any less real and true than the things they behave sensibly about?

Bernard Shaw

Contents

Preface

Quantum physics is the theory that underlies nearly all our current understanding of the physical universe. Since its invention some sixty years ago the scope of quantum theory has expanded to the point where the behaviour of subatomic particles, the properties of the atomic nucleus and the structure and properties of molecules and solids are all successfully described in quantum terms. Yet, ever since its beginning, quantum theory has been haunted by conceptual and philosophical problems which have made it hard to understand and difficult to accept.

As a student of physics some twenty-five years ago, one of the prime fascinations of the subject to me was the great conceptual leap quantum physics required us to make from our conventional ways of thinking about the physical world. As students we puzzled with this, encouraged to some extent by our teachers who were nevertheless more concerned to train us how to apply quantum ideas to the understanding of physical phenomena. At that time it was difficult to find books on the conceptual aspects of the subject – or at least any that discussed the problems in a reasonably accessible way. Some twenty years later when I had the opportunity of teaching Quantum Mechanics to undergraduate students, I tried to include some references to the conceptual aspects of the subject and, although there was by then a quite extensive literature, much of this was still rather technical and difficult for the non-specialist. With experience I have become convinced that it is possible to explain the conceptual problems of quantum physics without requiring either a thorough understanding of the wide areas of physics to which quantum theory has been applied or a great competence in the mathematical techniques that professionals find so useful. This book is my attempt to achieve this aim.

The first four chapters of the book set out the fundamental ideas of quantum physics and describe the two main conceptual problems: non-locality which means that different parts of a quantum system appear to influence each other even when they are a long way apart and even although there is no known interaction between them, and the 'measurement problem' which arises from the idea that quantum systems possess properties only when these are measured, although there is apparently nothing outside quantum physics to make the measurement. The later chapters describe the various solutions that have been proposed for these problems. Each of these in some way challenges our conventional view of the physical world and many of their implications are far-reaching and almost incredible. There is still no generally accepted consensus in this area and the final chapter summarizes the various points of view and sets out my personal position.

I should like to thank everyone who has helped me in the writing of this book. In particular Simon Capelin, Colin Gough and Chris Isham all read an early draft and offered many useful constructive criticisms. I was greatly stimulated by discussions with the audience of a class I gave under the auspices of the extra-mural department of the University of Birmingham, and I am particularly grateful for their suggestions on how to clarify the discussion of Bell's Theorem in Chapter 3. I should also like to offer particular thanks to Judy Astle who typed the manuscript and was patient and helpful with many changes and revisions.

Alastair I. M. Rae

see, quantum physics leads to the rejection of determinism – certainly of the simple type envisaged by Laplace – so that we have to come to terms with a universe whose present state is not simply 'the effect of its past' or 'the cause of its future'. Quantum theory tells us that nothing can be measured or observed without disturbing it, so that the role of the observer is crucial in understanding any physical process. So crucial in fact that some people have been led to believe that it is the observer's mind that is the only reality – that everything else including the whole physical universe is illusion. Others have suggested that quantum physics implies that ours is not the only physical universe, and that if we postulate the existence of myriads of universes with which we have only fleeting interactions, a form of determinism can be recovered. Others again think that, despite its manifest successes, quantum physics is not the final complete theory of the physical universe and that a further revolution in thought is needed. It is the aim of this book to describe these and other ideas and to explore their implications. Before we can do this, however, we must first find out what quantum physics is, so in this chapter we outline some of the reasons why the quantum theory is needed, describe the main ideas behind it, survey some of its successes and introduce the conceptual problems.

Light waves

Much of the evidence for the need for a new way of looking at things came out of a study of some of the properties of light. However, before we can discuss the new ideas, we must first acquire a little more detailed understanding of Maxwell's electromagnetic wave theory of light referred to earlier. Maxwell was able to show that at any point on a light beam there is an electric force and a magnetic force which are perpendicular both to each other and to the direction of the light beam. These forces (or *fields* as they are more properly called) oscillate many millions of times per second and vary periodically along the beam, as is illustrated in Figure 1.1. The presence of the electric fields could in principle be detected by measuring the electric voltage between two points across the beam. In the case of light such a direct measurement is quite impractical because the oscillation frequency is too large, but a similar measurement is actually made on radio waves every time they are received by an aerial on a radio or TV set. Direct evidence for the wave nature of light, on the other hand, is obtained from the phenomenon known as *interference*.

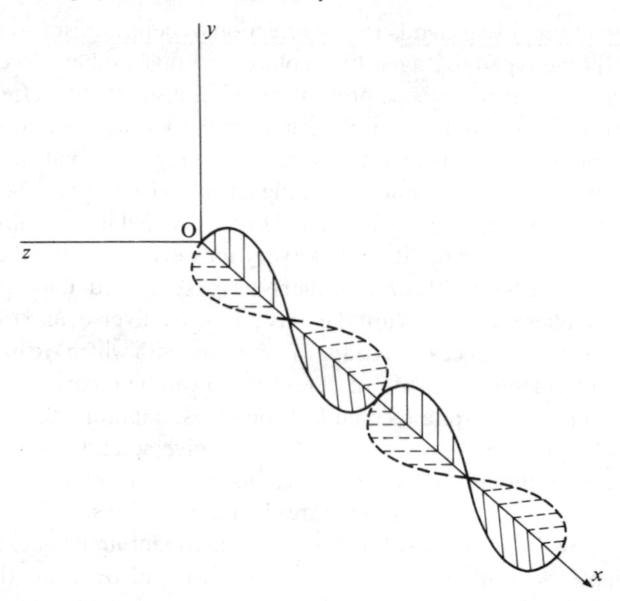

Fig. 1.1 An electromagnetic wave travelling along Ox consists of rapidly oscillating electric and magnetic fields which point along the directions Oy and Oz respectively.

An experiment to demonstrate interference is illustrated in Figure 1.2. Light passes through a narrow slit O after which it encounters a screen containing two slits A and B and finally reaches a third screen where it is observed. The light reaching the last screen can have travelled by one of the two routes – either by A or by B. But the distances travelled by light waves following these two paths are not equal and the light waves do not generally arrive at the screen 'in step' with each other. This point is also illustrated in Figure 1.2 from which we see that if the difference between the two path distances is a whole number of light wavelengths the waves reinforce each other, but if it is an odd number of half wavelengths they cancel each other out. As a result a series of visible light and dark bands is observed across the

Fig. 1.2 Light waves reaching a point on the screen C can have travelled via either of the two slits A and B. In (b) it is seen that if two paths differ by a whole number of wavelengths the waves add and reinforce, but if the path difference is an odd number of half wavelengths the waves cancel. As a result a series of light and dark bands are observed on the screen C in (a).

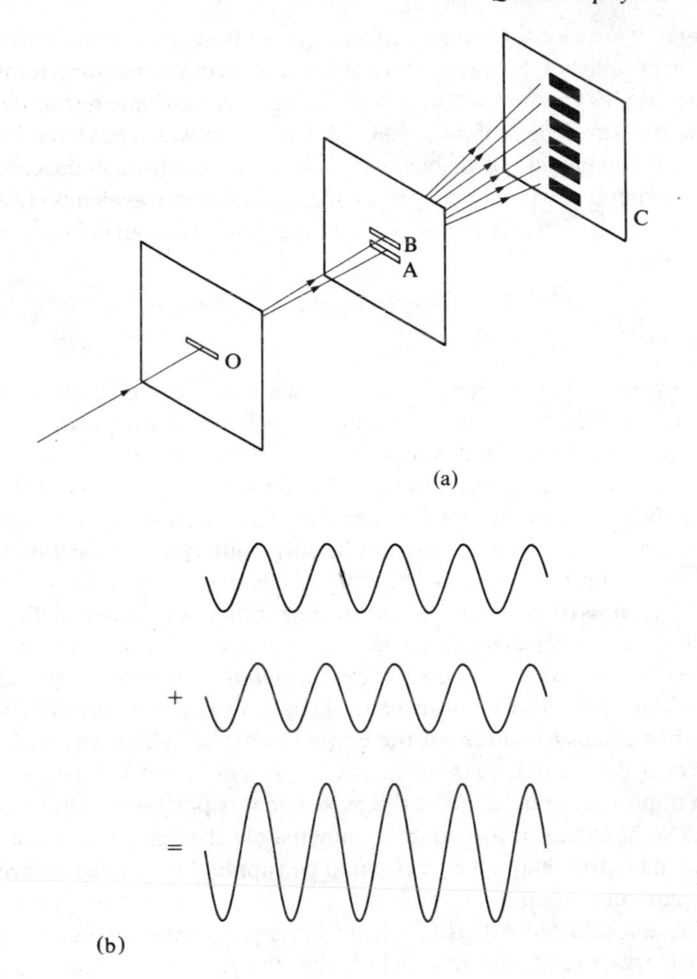

(a)

(b)

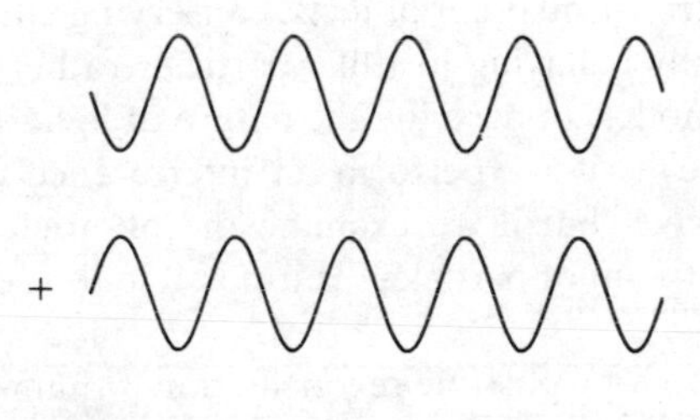

screen. It is the observation of effects such as these 'interference fringes' which establishes the wave nature of light. Moreover, measurements on these fringes can be used in a fairly straightforward manner to determine the wavelength of the light used, and in this way it has been found that the wavelength of visible light varies as we go through the colours of the rainbow, with violet light having the shortest wavelength (about 0.4 millionths of a metre) and red light the longest (about 0.7 millionths of a metre).

Photons

One of the first experiments to show that all was not well with classical physics (as the ideas of Newton and Maxwell are now termed) was the photoelectric effect. In this, light is directed onto a piece of metal in a vacuum and as a result subatomic charged particles known as electrons are knocked out of the metal and can be detected by applying a voltage between it and a collector plate. The surprising result of such investigations is that the energy of the emitted electrons does not depend on the brightness of the light, but only on its frequency or wavelength. For light of a given wavelength the *number* of electrons emitted per second increases with the light intensity, but the *amount of energy* acquired by each individual electron is unchanged. In fact the energy given to each electron equals $h\nu$ where ν is the frequency of the light wave and h is a universal constant of quantum physics known as Planck's constant. It is also important to note that if the experiment is performed with a very weak light electrons are emitted immediately the light is switched on and long before enough energy could be supplied by the light wave to any particular atom.

These results led Albert Einstein (the same scientist who developed the theory of relativity) to conclude that the energy in a light beam is carried in localized packets, sometimes known as 'quanta' or 'photons'. Further work has confirmed this and the photons have been seen to bounce off electrons and other objects, conserving energy and momentum and generally behaving just like particles rather than waves. We now have two models to describe the nature of light depending on the way we observe it: if we perform an interference experiment light behaves as a wave, but if we examine the photoelectric effect light behaves like a stream of particles. Is it possible to reconcile these two models?

One suggestion for a possible reconciliation is that we were mistaken ever to think of light as a wave: perhaps we should always have thought of it as a stream of particles with rather unusual properties which give rise to the interference patterns and mislead us into accepting the wave model. This would mean that the photons passing through the two slits of the apparatus shown in Figure 1.2 would somehow bump into each other, or at least interact in some way, so as to guide most of the photons into the light bands of the pattern and very few into the dark areas. This suggestion, although elaborate, is quite tenable in the context of most interference experiments because there is usually a large number of photons passing through the apparatus at any one time and interactions are always conceivable. If however we were to perform the experiment with very weak light, so that at any time there is only one photon in the region between the first slit and the screen, interactions between photons would be impossible and we must therefore expect the interference pattern to disappear. Such an experiment is a little difficult, but perfectly possible. The final screen must be replaced by a photographic plate or film and the apparatus must be carefully shielded from stray light, but if we do this and wait long enough for a large number of photons to have passed through one at a time, the interference pattern recorded on the photographic plate is just the same as it was before! We can go a little further and repeat the experiment several times using different lengths of exposure. The results of this are illustrated in Figure 1.3 from which we see that the photon nature of the light is

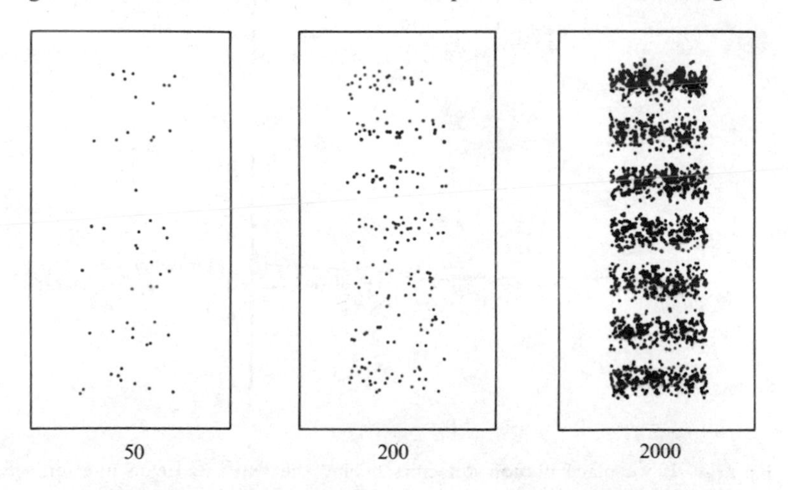

Fig. 1.3 The three panels show a computer reconstruction of the appearance of a two-slit interference pattern after 50, 200 and 2000 photons respectively arrive at the screen. The pattern appears clear only after a large number of photons have been recorded even though these pass through the apparatus one at a time.

confirmed by the appearance of individual spots on the photographic film. At very short exposures these seem to be scattered more or less at random, but the interference pattern becomes clearer as more and more arrive. We are therefore forced to the conclusion that interference does not result from interactions between photons. Indeed the fact that the interference pattern created after a long exposure to weak light is identical to one produced by the same number of photons arriving more or less together in a strong light beam implies that photons do not interact with each other at all.

Since interference does not result from interaction between photons, could it be that each individual photon somehow splits in two as it passes through the double slit? We could test for this if we put a photographic film or some kind of photon detector immediately behind the two slits instead of some distance away. In this way we can tell through which slit the photon passes, or if it splits in two on its way through (see Figure 1.4). If we do this, however, we always find that the photon has

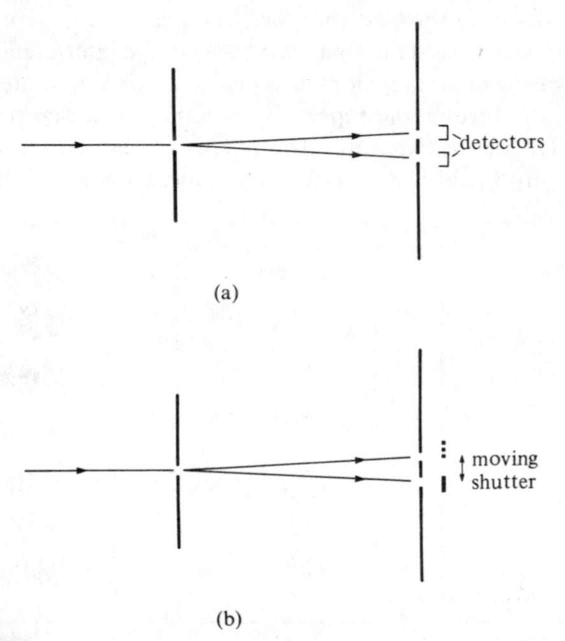

Fig. 1.4 If we place photon detectors behind the two slits of an interference apparatus, as in (a), each photon is always recorded as passing through one slit or the other and never through both simulaneously. If, as in (b), a shutter is placed behind the slits and oscillated up and down in such a way that both slits are never open simultaneously, the two-slit interference pattern is destroyed.

passed through one slit or the other and we never find any evidence for the photon splitting. Another test of this point is illustrated in Figure 1.4(b): if a shutter is placed behind the two slits and oscillated up and down so that only one of the two slits is open at any one time, the interference pattern is no longer formed. The same thing happens when any experiment is performed that detects, however subtly, through which slit the photon passes. It seems that light passes through one slit or the other in the form of photons if we set up an experiment to detect through which slit the photon passes, but passes through both slits in the form of a wave if we perform an interference experiment.

The fact that processes like two-slit interference require light to exhibit both particle and wave properties is known as *wave–particle duality*. It illustrates a general property of quantum physics which is that the nature of the model required to describe a system depends on the nature of the apparatus it is interacting with: light is a wave when passing through a pair of slits, but it is a stream of photons when it strikes a detector or a photographic film. It is this dependence of the properties of a quantum system on the nature of the observation being made on it that underlies the conceptual and philosophical problems which it is the purpose of this book to discuss. We shall begin this discussion in a more serious way in the next chapter, but we devote the rest of this chapter to a discussion of some further implications of the quantum theory and to an outline of some of its outstanding successes in explaining the behaviour of physical systems.

The Heisenberg uncertainty principle

One of the consequences of wave–particle duality is that it sets limits on the amount of information that can ever be obtained about a quantum system at any one time. Thus we can choose *either* to measure the wave properties of light by allowing it to pass through a double slit without detecting through which slit the photon passes *or* to observe the photons as they pass through the slits, so long as we sacrifice the possibility of performing an interference experiment, but we can never do both these things at the same time. Werner Heisenberg, one of the physicists who were instrumental in the early development of quantum physics, realized that this type of measurement and its limitations could be interpreted in a rather different way. He pointed out that the detection of which slit a photon went through was essentially a measurement of the position of the photon as it passes through the screen, while the observation of interference is akin to a measurement of momentum. It follows from wave–particle duality that it is impossible to make simultaneous position and momentum measurements on a quantum object such as a photon.

The application of Heisenberg's ideas to the two-slit experiment is actually rather subtle and a more straightforward example is the behaviour of light passing through a single slit of finite width. If this is analysed using the wave model of light, we find, as shown in Figure 1.5, that the slit spreads the light out into a 'diffraction pattern' whose angular spread is inversely proportional to the slit width. If we perform this experiment with very weak light so as to study the behaviour of individual photons we find that – just as in the two-slit experiment – the photons arrive at the screen more or less at random, and the diffraction pattern is built up gradually as more and more photons are accumulated. If we now look at this arrangement as a measurement of the position and momentum of the photon, we see that as the photon passes through the slit its position in the vertical direction in Figure 1.5 is determined by the size of the slit. What can we now say about the component of momentum in this direction? We know that when the photon arrives at the screen it will be found somewhere in the diffraction pattern, but we don't know where, and it therefore follows that the uncertainty in the vertical component of momentum is related to the angular spread of the pattern. Thus if we try to increase the accuracy of the position measurement by making the slit smaller, we will inevitably

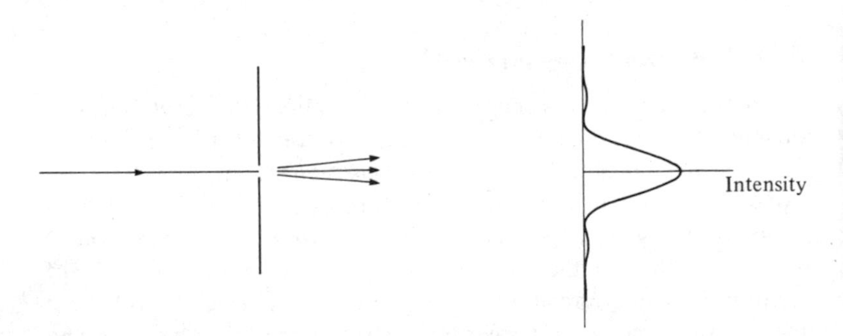

Fig. 1.5 Light passing through a single slit is diffracted to form a diffraction pattern whose intensity varies in the manner illustrated in the graph on the right. The narrower the slit, the broader is the diffraction pattern; as explained in the text, this result leads to limits on the possible accuracy of measurement of the position and momentum of the photons that are in accordance with Heisenberg's uncertainty principle.

increase the spread of the diffraction pattern and so reduce the precision of the momentum measurement. Heisenberg was able to show that quantum theory requires that all such measurements are subject to similar limitations. He expressed this in his famous *uncertainty principle* where the uncertainty (Δx) in position is related to that in momentum (Δp) by the relation

$$\Delta x \, \Delta p > h/4\pi$$

where h is the fundamental quantum constant (Planck's constant). If the single-slit diffraction pattern is analysed in more detail, the product of the position and momentum uncertainties can be shown to be about $\frac{1}{2}h$ which is clearly in agreement with Heisenberg's principle.

A common mistake in the application of the uncertainty principle to a situation like single-slit diffraction is to suggest that it is breached by the fact that the x component of momentum is known quite precisely when the photon arrives at the screen. The fallacy in this reasoning arises because when it is at the screen, the position of the photon is quite uncertain as it is no longer confined to the slit. It is the simultaneous determination of position and momentum whose precision is limited by the uncertainty principle.

The implications of the uncertainty principle on the way we think about scientific measurement are profound. It had long been realized that there are always practical limitations to the accuracy of any measurement, but before quantum physics there was no reason in principle why any desired accuracy should not be attainable by improving our experimental techniques. However, wave–particle duality and Heisenberg's uncertainty principle put a fundamental limit on the precision of any simultaneous measurement of two physical quantities such as the position and momentum of a photon. After this idea was put forward, there were a number of attempts to suggest experiments that might be able to make measurements more precisely than the uncertainty principle allows, but in every case careful analysis showed that this was impossible. It is now realized that the uncertainty principle is just one of the many strange and revolutionary consequences of quantum physics that have led to the conceptual and philosophical ideas that are the subject of this book.

Atoms and matter waves

Just as the wave model of light was well established in classical physics, there was little doubt by the beginning of the twentieth century that matter was made up of a large number of very small particles. Dalton's atomic theory had been remarkably successful in explaining chemical processes and the phenomenon of Brownian motion (in which smoke particles suspended in air are observed to undergo irregular fluctuations) had been explained as a result of the motion of discrete molecules. The study of the properties of electrical discharge tubes (the forerunners of the cathode-ray tube found in television sets) led J. J. Thompson to conclude that electrically charged particles (soon to be known as electrons) are emitted when a metal wire is heated to a high temperature in a vacuum. Very early in the twentieth century, Ernest Rutherford showed that the atom possessed a very small positively charged nucleus in which nearly all of the atomic mass was concentrated, and it was then easy to deduce that the atom must consist of this nucleus surrounded by electrons. At this point a problem arose. Every attempt to describe the structure of the atom in more detail using classical physics failed. The most obvious model was to suggest that the electrons orbit the nucleus as a planet orbits the sun, but Maxwell's electromagnetic theory requires that such an orbiting charge should radiate energy in the form of electromagnetic waves and, as this energy could come only from the electrons' motion, these would soon slow up and fall into the nucleus. The Danish physicist Niels Bohr, of whom we shall be hearing much more in due course, invented a model of the hydrogen atom (which contains only a single electron) in which such electron orbits were assumed to be stabilized under certain conditions and this model had considerable success. However, it failed to account for the properties of atoms containing more than one electron and there was no rationale for the rules determining the stability of the orbits.

At this point the French physicist Louis de Broglie put forward a radical hypothesis. If light waves sometimes behave like particles could it be that particles, such as electrons and nuclei, sometimes exhibit wave properties? To test such an apparently outrageous idea we might think of passing a beam of electrons through a two-slit apparatus of the type used to demonstrate interference between light waves (Figure 1.2). This is not possible because the wavelength predicted by de Broglie for such an electron beam is so short that the interference fringes would be too close together to be observed. However, shortly after de Broglie's suggestion, a very similar experiment was performed in which electrons were scattered off a crystal of nickel. An intensity pattern was observed which showed that interference had occurred between the electron waves scattered by different planes of atoms in the crystal and that the electron beam had indeed behaved like a wave in this situation. Much more recently it has been possible to produce neutron beams of wavelength comparable to that of light and these have been used to demonstrate two-slit interference in a manner very similar to the optical case confirming the presence of 'matter waves' in this situation too.

The matter-wave hypothesis was also confirmed indirectly, but possibly more dramatically, by its ability to explain the electronic structure of atoms. A proper understanding of this point requires a mathematical analysis that is well beyond the scope of this book, but the essence of the argument is that when waves are confined within a region of space only particular wavelengths are allowed: for example, only particular notes can be emitted by a violin string of given length and tension and similar principles govern the operation of most musical instruments. In an analogous way it is found that when the matter-wave hypothesis is combined with the fact that the negative electrons are attracted to the positive nucleus by an inverse square law of force, an equation results whose solutions determine the form of the electron waves in this situation. This equation (known as the Schrödinger equation after its inventor Erwin Schrödinger) has solutions for only particular 'quantized' values of the electron energy. It follows that an electron in an atom cannot have an energy lower than the lowest of these allowed values so the problem of the electron spiralling into the nucleus is avoided. Moreover, if an atom is 'excited' into an allowed state of energy higher than that of this lowest-energy 'ground' state it will jump back to the ground state while emitting a photon whose energy is equal to the difference between the energies of the two states (Figure 1.6). But

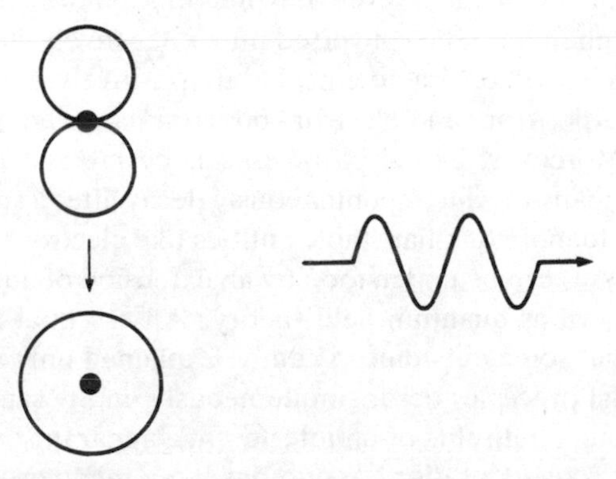

Fig. 1.6 Two of the possible stable patterns adopted for electron waves in atoms are shown on the left. If the atom makes a transition from the upper (higher-energy) to the lower state, a light photon of definite wavelength is emitted.

we have seen earlier that the energy of a photon is closely related to the wavelength of the associated light wave, so it follows that light is emitted by atoms at particular wavelengths only. It had been known for some time that light emitted from atoms (in gas-discharge tubes for example) had this property and it is a major triumph of quantum physics that this can be explained not just qualitatively, but also by detailed calculations of the allowed wavelengths which are found to be in excellent agreement with experiment.

Beyond the atom

The success of the matter-wave model did not stop at the atom. Similar ideas were applied to the structure of the nucleus itself which is known to contain an assemblage of positively charged particles, called protons, along with an approximately equal number of uncharged neutrons. The form of the force between these particles is not known precisely and it is much more complex than an inverse square law, so the calculations are considerably more difficult than in the atomic case. The results, however, are just as good and the calculated properties of atomic nuclei are also found to be in excellent agreement with experiment.

Nowadays even 'fundamental' particles such as the proton and neutron (but not the electron) are known to have a structure and to be composed of even more fundamental objects known as 'quarks'. This structure has also been successfully analysed by quantum physics in a similar manner to those of the nucleus and the atom, showing that the quarks also possess wave properties. But modern particle physics has extended quantum ideas even beyond this point. At high enough energies a photon can be converted into a negatively charged electron along with an otherwise identical, but positively charged, particle known as a positron, and electron–positron pairs can recombine into photons. Moreover, exotic particles can be created in high-energy processes, many of which spontaneously decay after a small fraction of a second into more familiar stable entities like electrons or quarks. All such processes can be understood by an extension of quantum ideas in a form known as quantum field theory. An essential feature of this theory is that some phenomena can be explained only if a number of fundamental processes occur simultaneously: in the same way as light passes through both slits of an interference apparatus, even though it apparently consists of discrete photons, so a number of fundamental quantum-field processes add together in a coherent way to create the observed phenomenon.

Condensed matter

The successes of quantum physics are not confined to atomic or subatomic phenomena. Soon after the establishment of the matter-wave hypothesis, it became apparent that it could also be used to explain chemical bonding. For example, in the case of a molecule consisting of two hydrogen atoms, the electron waves surround both nuclei and draw them together with a force that is balanced by the mutual electrical repulsion of the positive nuclei to form the hydrogen molecule. These ideas can be developed into calculations of molecular properties, such as the equilibrium nuclear separation, which agree precisely with experiment. The application of similar principles to the structure of condensed matter, particularly solids, has been just as successful. Quantum physics can be shown to account for the fact that some solids are insulators while others are metals that conduct electricity and others again – notably silicon and germanium – are semiconductors. The special properties of silicon that allow the construction of the silicon chip with all its ramifications turn out to be direct results of the existence of electron waves in solids. Even the exotic properties of materials at very low temperatures, where liquid helium has zero viscosity and some metals become superconductors completely devoid of electrical resistance, can be shown to be manifestations of quantum behaviour.

The last three sections of this chapter only touch on some of the manifest successes quantum physics has achieved over the last half-century. Wherever it has been possible to perform a quantum calculation of a physical quantity it has always been in excellent agreement with the results of experiment. However, the purpose of this book is not to survey this achievement in detail, but rather to explore the fundamental features of the quantum approach and to explain their revolutionary implications for our conceptual and philosophical understanding of the physical world. To achieve this we need a rather more detailed understanding of quantum ideas than we have obtained so far and we begin this task in the next chapter.

2 · Which way are the photons pointing?

The previous chapter has surveyed part of the rich variety of physical phenomena that can be understood using the ideas of quantum physics. Now that we are beginning the task of looking more deeply into the subject we shall find it very useful to concentrate on examples that are comparatively simple to understand, but which still illustrate the fundamental principles and highlight the basic conceptual problems. A few years ago most writers discussing such topics would have naturally turned to the example of the 'particle' passing through the two-slit apparatus, with its wave properties being revealed in the interference pattern (as in Figure 1.2), and much of the discussion would have been in terms of 'wave–particle duality'. Nowadays, however, it is realized that there are considerable advantages in concentrating on situations where there are only a small number of possible outcomes of a measurement. For this reason we shall describe a further property of light that may not be familiar to all readers. This is known as *polarization*. In the next section we discuss it in the context of the classical wave theory of light, while the rest of the chapter extends the concept to situations where the photon nature of light is important.

The polarization of light

Imagine that a beam of light is coming towards us and that we think of it as an electromagnetic wave. As we saw in Chapter 1 this means that at any point in space along the wave there is an electric field which is vibrating many times per second. At any moment in time, this electric field must be pointing in some direction, and it turns out that Maxwell's equations require the direction of vibration always to be at right angles to the direction of travel of the light. So if the light is coming towards us the electric field may point to the left or the right or up or down or in some direction in between, but not towards or away from us (Figure 2.1). In many cases the electric field direction changes rapidly

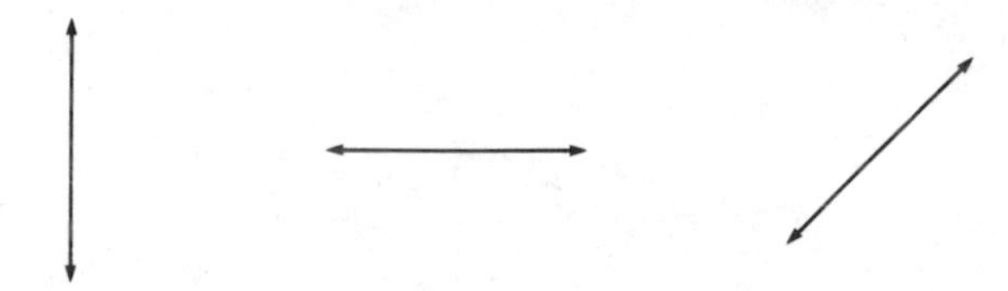

Fig. 2.1 For a light wave coming towards us the electric field may oscillate vertically, horizontally or at some angle in between, but the oscillation is always perpendicular to the direction of the light beam.

from time to time, but it is possible to create light where this pointing direction remains constant. Such light is said to be *plane polarized* (because the electric field is always in the same plane) or sometimes just *polarized*. The plane containing the electric field vectors is known as the *plane of polarization* and the direction in which the electric field points is known as the *polarization direction*.

The idea of polarization may be more familiar to some readers in the context of radio or T.V. reception. To get a good signal into a receiver it is necessary to align the aerial dipole along the polarization direction (usually horizontal or vertical) of the radio waves. This ensures that the electric field will drive a current along the aerial wire and hence into the set.

Polarized light can be produced in a number of ways. For example light from most lasers is polarized as a result of internal processes in the laser, but a polarized beam is most conveniently produced from any beam of light using a substance known as 'polaroid'. This substance is actually a thin film of nitrocellulose packed with extremely small crystals, but the construction and operational details are not relevant to our discussion. What is important is that polaroid seems to act by 'filtering out' all but the required polarization: by this we mean that if ordinary unpolarized light shines on one side of a piece of polaroid, the light emerging from the other side is polarized and has an intensity about half that of the incident light (Figure 2.2). The polarization direction of the light coming out of a polaroid is always along a particular direction in the polaroid sheet which we call the polaroid axis.

Polaroid can also be used to find the polarization direction of light that is already polarized: we just turn the polaroid round until the emerging light is as strong as the light that went in. A very importantt point to note is that the polaroid axis does not have to be exactly lined up with the polarization direction before any light comes through at all. The light is stopped completely only if the two axes are at right angles,

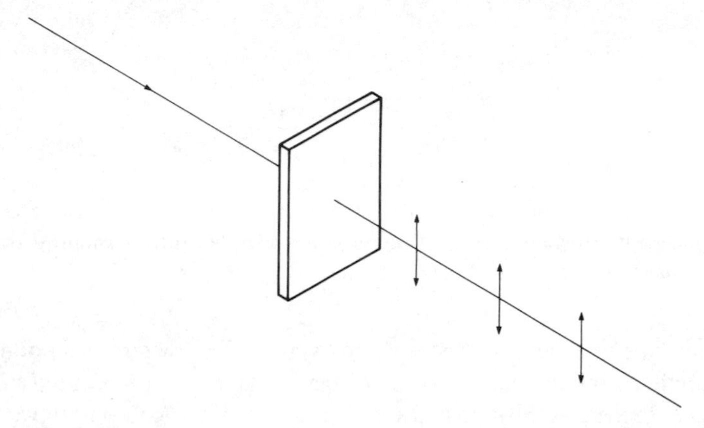

Fig. 2.2 If a light beam passes through a piece of polaroid the electric vector of the emitted light is always parallel to a particular direction (vertical in the case shown) known as the polaroid axis.

and the transmitted fraction increases gradually and smoothly as the polaroid is rotated until maximum transmission occurs when the axes are parallel. In slightly more technical language, we say that the polaroid allows through the 'component' of the light which is polarized in the direction of the polaroid axis. This is illustrated in Figure 2.3 which shows how an electric field in a general direction (OP) can be thought of as the addition of two components (OA and OB) at right angles to each other. If the polaroid axis points along OA, say, the direction and magnitude of the electric field of the emergent light are

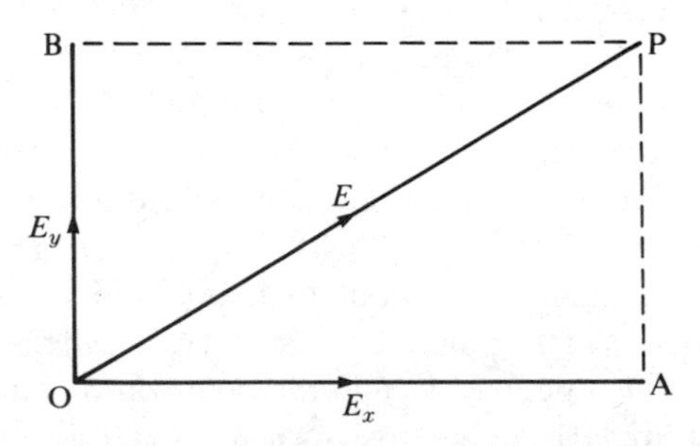

Fig. 2.3 A vibration along the direction OP can be thought of as a combination of vibrations along OA and OB. If light with electric field amplitude OP (intensity OP^2) is passed through a polaroid with its axis along OA, the amplitude of the transmitted light will be equal to OA and its intensity will be OA^2.

the same as those of this component of the incident light. Similar results are obtained for other orientations of the polaroid by redrawing Figure 2.3 with the angle POA changed appropriately; the particular cases POA = 0° and POA = 90°, clearly correspond to transmission of all and none of the light respectively and, in general, the intensity of the emergent light is equal to $E_x^2 = I \cos^2 \theta$ where $I = E^2$ is the intensity of the incident intensity and θ is the angle POA.

A slightly different kind of device used for generating and analysing the polarization of light is a single crystal of the mineral calcite. The details of operation need not concern us, but such a crystal is able to separate the light into two components of perpendicular polarization (Figure 2.4). Unlike polaroid which absorbs the component that is perpendicular to the polaroid direction, the calcite crystal allows all the light through, but the two components emerge along different paths. We see that the total light emerging in the two beams is equal to that entering the calcite crystal because the intensity of the light is proportional to the square of the electric field and, referring to Figure 2.3, the incident intensity, E^2, equals the sum of the two transmitted intensities E_x^2 and E_y^2 by Pythagoras' Theorem.

Consideration of the analysis of a light beam into two polarization components by a device such as a calcite crystal will play a central role in much of the discussion in later chapters. However, as the details of how this is achieved are not of importance, from now on we shall illustrate the process simply by drawing a square box, as in Figure 2.5, with one incident and two emergent light beams. The label 'HV' on the box shows that it is oriented so that the emergent beams are polarized in the horizontal and vertical directions; we shall also consider other orientations such as '±45°' which means that the emergent beams are polarized at +45° and −45° to the horizontal respectively.

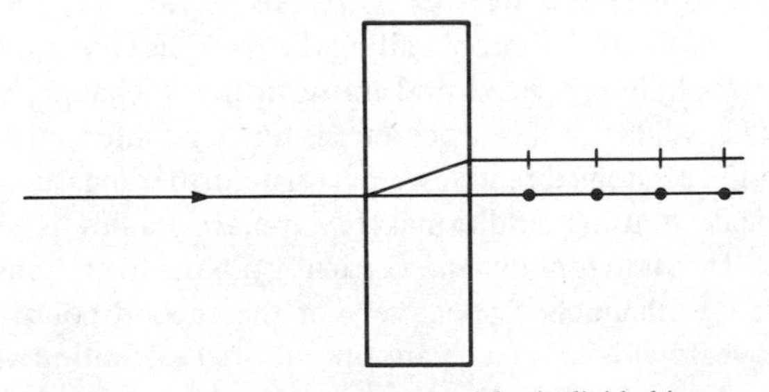

Fig. 2.4 Light passing through a crystal of calcite is divided into two components whose polarizations are respectively parallel and perpendicular to a particular direction in the calcite crystal.

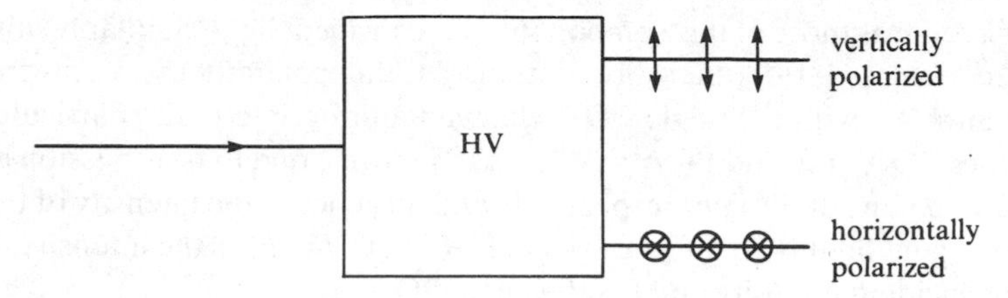

Fig. 2.5 In discussing polarization, we represent a polarizer (such as that illustrated in Figure 2.4) by a box with a legend indicating the direction of the polarization axis. In the example shown the box resolves the incident light into components polarized in the vertical and horizontal directions.

The polarization of photons

We saw in the last chapter that the classical wave theory is unable to provide a complete description of all the properties of light. In particular, when light is detected by a device based on the photoelectric effect, it behaves as if it consisted of a stream of particles, known as photons. The photon nature of light is particularly noticeable if the overall intensity is very low so that its arrival at the detector is indicated by occasional clicks: at higher intensities the clicks run into each other and the behaviour is the same as would be expected from a continuous wave. At first sight it might seem that polarization is very much a wave concept and might not be applicable to individual photons. However, this is not so as can be seen easily by considering the experiment illustrated in Figure 2.5: if a very weak beam of unpolarized light is incident on a polarizer and the two output beams are directed onto detectors capable of counting individual photons, the photons must emerge in one or other of the two channels (if only because they have nowhere else to go!) and we can therefore ascribe the polarization property to an individual photon, calling the photons emerging in the H channel horizontally polarized and those in the V channel vertically polarized. The validity of this procedure is further confirmed if, instead of detecting the photons directly, we perform further measurements of their polarization using additional HV polarizers, as is shown in Figure 2.6: all the horizontally and vertically polarized photons emerge in the H and V channels respectively of the second polarizers. We therefore have what is known as an 'operational' definition of photon polarization. That is, although we may not be able to say what it is, we can define it in terms of the operations that have to be performed in order to measure it. Thus a horizontally polarized photon, for example,

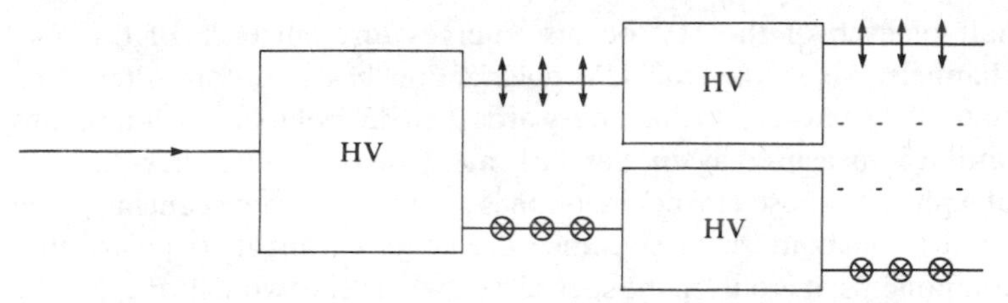

Fig. 2.6 Polarization is a property that can be attributed to photons because every photon that emerges from the first polarizer as vertically or horizontally polarized passes through the same channel of the subsequent HV polarizer.

is one that has emerged through the horizontal channel of an HV–oriented calcite crystal or through a polaroid whose axis is in the horizontal direction.

We can now see how a consideration of the measurement of photon polarization illustrates some of the general features of quantum measurement. We saw in the previous chapter how a measurement of some physical property, such as a particle's position, could not be made without disturbing the system so that the results of measurement of some other property, such as the particle's momentum, become unpredictable. Consider now the set-up shown in Figure 2.7. A beam of photons, all polarized at 45° to the horizontal (obtained perhaps by passing an unpolarized light beam through an appropriately oriented polaroid) is incident on a polarizer oriented to measure HV polarization. Half the photons emerge in the horizontal channel and the other half in the vertical channel. If we now pass either beam through another polarizer oriented to perform ±45° measurements, we find that the original 45° polarization has been destroyed by the HV measurement:

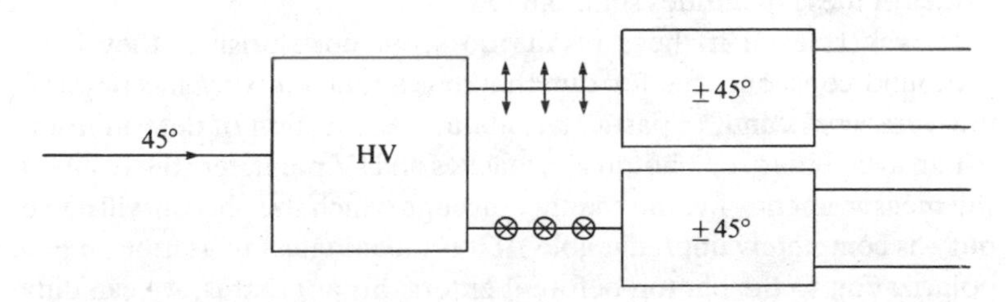

Fig. 2.7 If 45° polarized photons are incident on an HV polarizer they emerge as either horizontally or vertically polarized. They pass at random through the two channels of the later ±45° polarizers showing that they have lost their 'memory' of their original polarization. We conclude that polarization measurements generally change the polarization of the measured photons.

half of each of the HV beams emerges through each of the ±45° channels. Measuring the HV polarization has therefore altered the state of the photons so that afterwards their 45° polarization is unknown until it is measured again. Similarly a 45° measurement alters the state of a photon whose HV polarization is known. The same conclusion can be drawn about successive measurements of any two polarization components: it is only in the special case where the two sets of apparatus are parallel (or perpendicular) that a second measurement can be made without disturbing a previously known polarization state of the photons.

When applied to photon polarization, results such as those described in the previous paragraph may seem quite unsurprising. After all when we thought of light as an electromagnetic wave, we saw that the effect of a polarizer was, in general, to split the wave into two components neither of which is the same as the incident wave. Moreover, whereas we can say that a horizontally polarized wave definitely has horizontal polarization and is certainly not vertically polarized, we cannot say *at the same time* that it is polarized at +45° to the horizontal and not at −45° or vice versa. Our conclusions about photon polarization (confirmed by experiment) followed directly from an extension of the ideas based on the wave model to the case of very weak light, so we should not be surprised that it is similarly impossible to attribute simultaneously different polarizations to a photon or that a measurement of one photon polarization changes the state of the photon and destroys its previous polarization. Indeed we might generalize this idea to other quantum measurements and suggest that if, for example, we understood properly what the concepts of position and momentum mean on an atomic scale we might find it equally illogical to expect a particle to possess definite values of these quantities simultaneously.

However, even if these results do seem unsurprising, they have profound consequences for our thinking about the way the physical universe works and, in particular, about the question of determinism. When a 45° polarized photon approaches an HV polarizer, the result of the measurement – i.e. the channel through which the photon will come out – is completely unpredictable. If it is meaningless to ascribe an HV polarization to the photon before it enters this apparatus, we can only conclude that the outcome of the measurement is determined purely by chance. We know that if there are many photons, then on average half of them will appear in each channel, but if we concentrate on one photon only, its behaviour is completely random. Indeed (although we must be careful about using such language) we could say that even the photon does not know through which channel it is going to emerge! We should note that this unpredictability is a result of the presence of photons, and would not arise if the wave model were completely true: an HV polarizer allows half of the incoming 45° polarized wave to pass along each channel and we don't need to worry about which half; it is only because the photon apparently has to pass through either one channel or the other that the indeterminacy arises. However, although the indeterminacy results from the behaviour of the photon, it is important to realize that it doesn't stop there. If we consider a single photon passing through the apparatus shown in Figure 2.7, the fact that it emerges at random through one or other channel means that we cannot predict which of the two detectors will record an event. But the operation of a detector is not a microscopic event, it operates on the large scale of the laboratory or of everyday events. Indeed there is no reason why the operation of the detectors should not be coupled to some more dramatic happening, such as the flashing of a light or the raising of a flag (Figure 2.8). If we consider a single photon entering a set-up like this, it is quite unpredictable whether or not the light will flash or the flag will run up the flagpole!

As we pointed out in Chapter 1, the conclusion of quantum physics that some events, at least, are essentially unpredictable is completely contradictory to the classical view of physics which maintains that the behaviour of the universe is governed by mechanical laws. Classically particles were thought to move under the influence of definite forces and if all these forces were known, along with the positions and speeds of all the particles at some instant, the subsequent behaviour of any physical system could be predicted. Of course such calculations would be practicable only in simple cases, but in principle it would be possible to predict the behaviour of any physical system – even the whole

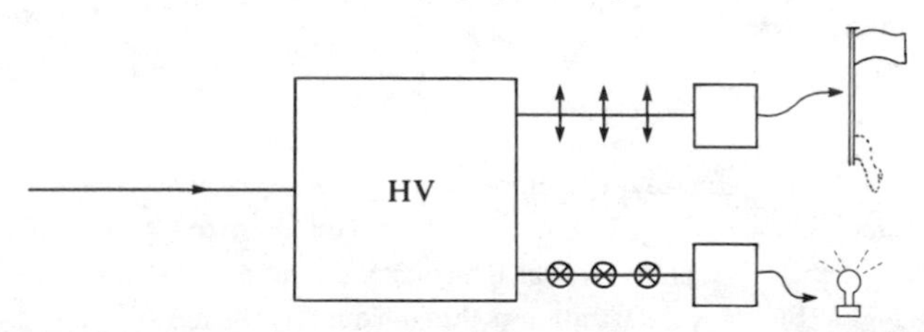

Fig. 2.8 The channel through which a photon will emerge is determined purely by chance. If an outgoing vertically polarized photon triggers a device to raise a flag while an outgoing horizontal photon causes a light to be switched on, then which of these two events occurs for a given incident photon is completely unpredictable and undetermined.

universe. Quantum physics has destroyed this deterministic view, and we see that indeterminism and uncertainty are built into the very foundations of the theory. In general the future behaviour of a physical system cannot be predicted however accurately its present state is known.

We close this section by introducing a variant on the photon-polarization measurement that illustrates the key problem of quantum measurement and to which we shall return frequently in later chapters. As shown in Figure 2.9, a 45° polarized light beam is split into its horizontal and vertical components by an HV polarizer in the same way as before, and the two light beams are directed onto a second calcite crystal that faces in the opposite direction. We first consider the case

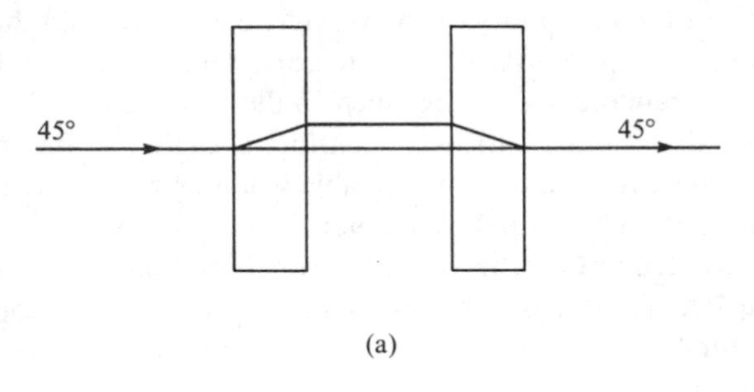

(a)

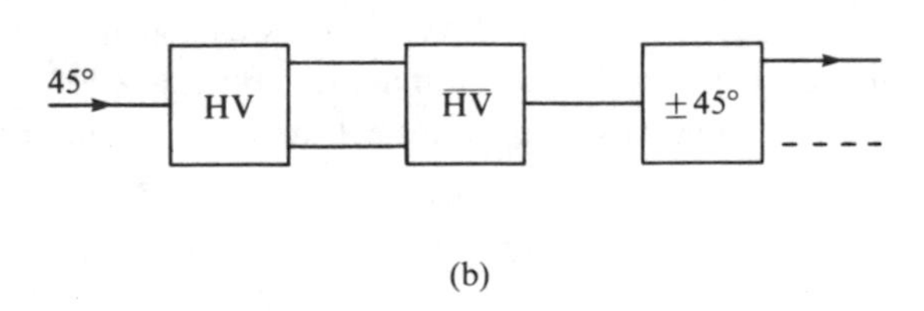

(b)

Fig. 2.9 (a) Light split into two components by a calcite crystal can be reunited by a second calcite crystal facing in the opposite direction. If the crystals are set up carefully so that the two paths through the apparatus are identical the light emerging on the right has the same polarization as that incident on the left.
(b) The boxes HV and $\overline{HV}$ represent forward and reversed polarizers as in (a). It follows that a 45° polarized light will still have this polarization when it emerges from the box $\overline{HV}$. This is also true for individual photons, a fact which is difficult to reconcile with the idea of measurement changing the polarization state (cf. Figure 2.7).

where the light intensity is so high that we can apply the wave model with confidence. One of the properties of polarized waves in calcite crystals is that, as a result of the wavelength of the beam going straight through being a little shorter than that of the other, the two light waves emerging from the first crystal have performed an identical number of oscillations and are therefore 'in step'. The same thing happens in the second crystal and the net result is that the polarizations of the two components emerging together from this add and recreate a 45° polarized wave. It should be noted that, although such an experiment requires careful setting up to ensure that the above conditions are accurately fulfilled, it is a real experiment that can be and often is performed in a modern laboratory.

Now suppose that the light entering the apparatus is so weak that only one photon passes through at any one time. From what we already know we might be tempted to reason as follows: earlier experiments have shown us that a 45° photon follows only one of the two possible paths through the HV apparatus; when it reaches the right-hand crystal there is therefore nothing with which it can be reunited so there is no way in which its 45° polarization can be reconstructed and it will emerge at random through either channel of the final 45° polarizer. When the experiment is performed, however, *the opposite result is observed*: however weak the light, the emergent beam is found to have 45° polarization, just as the incident beam has. In this case the light behaves just as if each photon had split and followed both paths to be reunited in the crystal. Yet we know that if, instead of the second crystal, we had inserted two detectors, as in Figure 2.8, we would have found that the photon was certainly in one channel or the other and not both. Moreover there is no reason why the reunifying crystal and the detectors should not be a long way from the original polarizer – at the other side of the laboratory or even, in principle, the other side of the world. The light entering the HV polarizer can then have no knowledge of what kind of apparatus it is going to encounter on the other side and the possibility of it adjusting its behaviour in the polarizer so as to suit the subsequent measurement is eliminated.

A further possibility may have occurred to some readers. Could it be that the effect of the second crystal is different from what we thought and that really any horizontally or vertically polarized photon is turned into a 45° photon by such an apparatus? This idea could be tested by blocking off one or other of the HV beams so that the light entering the crystal is certainly either horizontally or vertically polarized. But, whenever this is done, we find that the initial polarization has indeed been destroyed and the photons emerge at random through the two channels of the final 45° polarizer. We seem therefore to be forced to the conclusion that the photon either passes along both HV channels at once (despite the fact that we can only ever detect it in one channel) or, if it passes along only one path, it somehow 'knows' what it would have done if it had followed the other! This is another example of the strange consequences of quantum theory that we encountered in our discussion of two-slit interference in the last chapter.

Hidden variables

The strange, almost paradoxical, nature of quantum phenomena such as those described above has led some workers to attempt to devise models of the physics of subatomic particles that would account for such observations in a more sensible, rational way. Theories of this kind are known as 'hidden-variable theories' for reasons that should become clear shortly.

A simple hidden-variable theory that could account for the apparent indeterminism of the outcome of the simple polarization measurement (Figure 2.7) would be to attribute to the 45° photon some property which decided in advance through which HV channel it would pass. Thus our statement that the photon 'did not know' which way it was going to go would be wrong: although we could not measure it, this 'hidden' property would determine the outcome of the HV experiment. We can draw an analogy to the tossing of a coin: although in practice we cannot tell in advance whether it will come down heads or tails the result is determined by the laws of classical physics along with the initial speed and spin of the coin. The difference in the quantum case is that any corresponding properties determining which way the photon will go are unmeasurable *in principle*: they are 'hidden variables'.

Although a simple hidden-variable theory, like that just described, could account for indeterminism, it cannot readily explain the reunification experiment discussed at the end of the previous section and illustrated in Figure 2.9. We still have the problem of how the photon knows what is happening in the channel it is not passing through. To overcome this a slightly different form of wave–particle duality has been suggested. Instead of treating the wave and particle models as alternatives, this theory proposes that both are present simultaneously in a quantum situation. The wave is no longer directly detectable, as the electromagnetic wave was thought to be, but has the function of guiding the photon along and adjusting its polarization. Thus the 45° polarized photon approaches the HV polarizer and sees the guiding wave being split into two parts. It follows one or other path (probably at random because this form of hidden-variable theory does not maintain determinism) and has its polarization adjusted to fit the wave in that path. The same thing happens when it passes through the reunifying crystal: the outgoing wave has 45° polarization (because it is an addition of the H and V waves in step) and this property is transmitted to the outgoing photon. This model therefore preserves *locality*, by which we mean that the behaviour of the photon results from the properties of the wave at the point where the photon happens to be. This is in contrast to the conventional quantum theory where the photon appears to be affected by the whole apparatus including the path it isn't following.

Hidden-variable theories, such as those just described, have been suggested by a number of workers, notably the physicist David Bohm. They have been developed to the stage that most of the results of conventional quantum physics can be reproduced by a theory of this kind. However the theories have their own problems which many people think are just as conceptually unacceptable as those of the conventional quantum approach. In particular, the mathematical details of hidden-variable theories are much more complex than those of quantum physics, which are basically simple and elegant; the 'guiding wave' seems to be quite unlike any other wave field known to physics: it possesses no energy of its own yet it is able to influence the behaviour of its associated particles. But hidden-variable theories have a further, and some believe fatal, disadvantage. Although designed to preserve locality in situations such as those discussed, it turns out that they are unable to do so in all circumstances. In particular, some situations involving the quantum behaviour of pairs of photons turn out to be inexplicable using any local hidden-variable theory. This would seem to rule out the main advantage of a hidden-variable theory, but when this was first suggested it was realized that no experimental tests of the correlated behaviour of photon pairs had been carried out, so it was possible that quantum physics is actually wrong in such situations and some form of local hidden-variable theory is correct. This possibility has been the subject of considerable theoretical and experimental investigation over the past few years and, because of its importance and the light it sheds on our general understanding of quantum phenomena, the next chapter is devoted to a reasonably detailed explanation of this work.

3 · What can be hidden in a pair of photons?

Albert Einstein's comment that 'God does not play dice' sums up the way many people react when they first encounter the ideas discussed in the previous chapters. How can it be that future events are not completely determined by the way things are at present? How can a cause have two or more possible effects? If the choice of future events is not determined by natural laws does it mean that some supernatural force (God?) is involved wherever a quantum event occurs? Questions of this kind trouble most students of physics, but nearly all, conditioned by a scientific education, get used to the conceptual problems, say 'Nature is just like that' and apply the ideas of quantum physics to their study or research without worrying about their fundamental truth or falsity. Some physicists, however, never get used to the, at least apparent, contradictions and believe that the fundamental physical processes underlying the basic physics of the universe must be describable in deterministic, or at least objective realistic, terms. Einstein was one of those. He stood out obstinately against the growing consensus of opinion in the nineteen-twenties and -thirties that was prepared to accept indeterminism and the lack of objective realism as a price to be paid for a theory which was proving so successful in a wide variety of practical situations. Ironically, however, Einstein's greatest contribution to the field was not some subtle explanation of the underlying structure of quantum physics, but the exposure of an even more astonishing consequence of quantum theory. This arises from the analysis of the quantum behaviour of systems containing two or more particles that interact and move apart. Einstein was able to show that, in some circumstances, quantum physics implies that the separated particles influence each other even when there is no known interaction between them. The arguments underlying these conclusions are discussed in this chapter; unfortunately they are unavoidably rather more complex and technical than those in the rest of this book, but they are very important and worth the effort that may be required.

The ideas to be discussed in this chapter were first put forward by Einstein and his co-workers, Boris Podolski and Nathan Rosen, in 1935 and because of this the topic is often denoted by their initials EPR. Their arguments were presented in the context of wave–particle duality, but in 1951 David Bohm showed that the point could be made much more clearly if we consider the measurement of variables, like photon polarization*, whose results were limited to a small number of possible values. To understand the EPR problem we consider a physical system consisting of atoms in which a transition occurs from an excited state to the ground state with the emission of two photons in rapid succession. The wavelengths of the two photons are different, so they correspond to two different colours, say red and green, but their most important property is that their polarizations are always at right angles: if the red photon is vertically plane polarized then the green photon is horizontally plane polarized or if one is polarized at +45° to the horizontal the other is at −45° and so on. Of course not all atoms that emit photons in pairs have this property, but some do and experiments on such systems are perfectly practicable as we shall see.

How do we know that the polarizations are always at right angles? One answer might be that the quantum theory of the atom requires it, but a more important reason is that this property can be directly measured. Consider the arrangement shown in Figure 3.1. A gas of atoms that emit pairs of photons is placed between two filters, one of

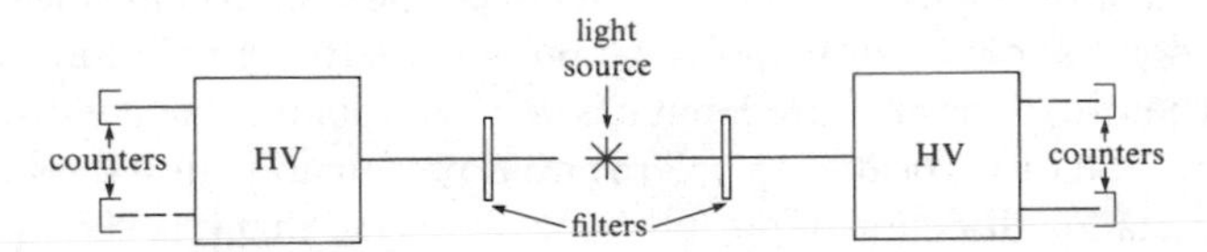

Fig. 3.1 In some circumstances atoms can be made to emit a pair of photons in rapid succession. The two members of each pair move away from the source in opposite directions; as they have different wavelengths they can be identified by passing the light through appropriate filters. In the set-up shown the HV polarization of one of the two photons is measured by the right-hand apparatus and that of the other is measured on the left. Whenever a right-hand photon is found to be horizontally polarized that on the left is recorded as vertical and *vice versa*.

* Bohm's paper actually related to the measurement of the angular momentum or 'spin' of atoms. However, it turns out that both the experimental measurements of and the theoretical predictions about particle spin are practically identical to those relating to photon polarization.

which allows through red light and the other green light. Each of these light beams is directed at an HV polarizer of the type described in the last chapter and the outputs from the two channels of each polarizer are monitored by photon detectors. The intensity of the emitted light is arranged to be sufficiently low and the detectors operate fast enough for the pairs of photons to be individually detected. When the apparatus is switched on it is found that every time a photon is detected in the H channel of the left-hand polarizer, a V photon is found on the right and *vice versa*. But of course there is nothing special about the HV configuration and if both polarizers are rotated through the same arbitrary angle, the same result is obtained: for example if they are set up to make 45° measurements, a left-hand +45° photon is always accompanied by a right-hand −45° photon and *vice versa*.

This may seem quite straightforward, but now consider the set-up in Figure 3.2. This is just the same as that considered before except that we have removed the right-hand polarizer and detectors because they are unnecessary! If the right-hand polarization is always perpendicular to that on the left then we know what the right-hand polarization is without measuring it. Or, putting it another way, the act of measuring the left-hand polarization also constitutes a measurement of the polarization of the right-hand photon. But just a minute! This would be perfectly all right if we were making a conventional classical measurement, but we saw in the previous chapter that an important feature of any quantum measurement is that it affects the system being measured. We don't know what the polarization of the left-hand photon actually was before it was measured, but it is very improbable that it was exactly horizontal or vertical. The polarizer has presumably turned round the polarization direction of the left-hand photon so that it is H or V but it obviously cannot have affected the right-hand photon which is on the other side of the laboratory by the time the left-hand measurement is made!

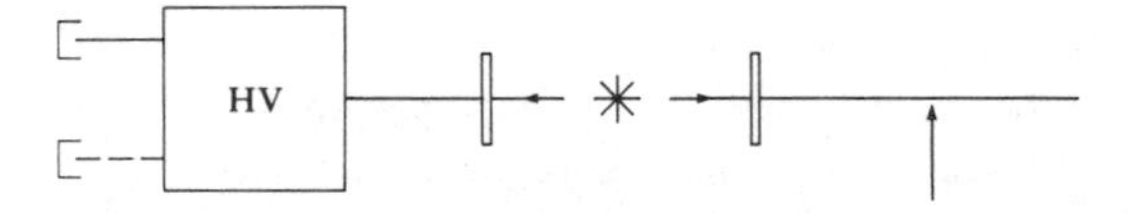

Fig. 3.2 Since we know the polarizations of the pair of photons are always at right angles, the right-hand apparatus of Fig. 3.1 is unnecessary. Whenever a left-hand photon is detected as vertical we can conclude that its partner (passing the point arrowed) must be horizontal and *vice versa*. But if a quantum measurement alters the state of the object measured how can the left-hand apparatus affect the polarization of the right-hand photon many metres away?

If we reject the idea of the measuring apparatus affecting a photon at a distance, how might we explain the results described above? One possible idea would be to suggest that the effect described in Figure 3.1 depends crucially on the presence of *both* polarizers and associated detectors. After all the polarization of the right-hand photon in Figure 3.2 is not measured, so we have no means of knowing that it is really polarized at right angles to the left-hand photon, and it might well be the insertion of the right-hand polarizer and detector which produces the effect. We could simply postulate that each photon interacts with the measuring apparatus in the same way so that, whatever their actual polarization before they are detected, they always emerge in opposite channels of the two sets of apparatus. In saying this, however, we have implicitly rejected the quantum idea of the measurement being a random, indeterministic process. The photons are *always* detected with perpendicular polarization so if this results from their interactions with the two sets of measurement apparatus there can be no room for any randomness associated with this interaction. In other words we seem to have strong evidence for a deterministic hidden-variable theory, of the type discussed in Chapter 2: the outcome of the photon-polarization measurement appears to be determined in advance by some property of the photon – each photon 'knows what it is going to do' before it enters the polarizer.

Such a conclusion is similar to that reached by Albert Einstein and his co-workers in their original paper whose title is 'Can quantum-mechanical description of physical reality be considered complete?' If the left-hand apparatus cannot affect the state of the right-hand photon, then the set-up in Figure 3.2 must measure some property of the right-hand photon without disturbing it. Even if this property is not the polarization itself it must be related to the hidden variable which would determine the result of a polarization measurement. This quantity must therefore be 'real'. As Einstein puts it –

> If, without in any way disturbing the system, we can predict with certainty (i.e. with probability equal to unity) the value of a physical quantity, then there exists an element of physical reality corresponding to this physical quantity.

We therefore seem to be left with a choice: either the ideas of quantum measurement can be extended so that an apparatus affects a photon a long way from it, or there is a deterministic hidden-variable theory underlying quantum physics. What we now want is an experiment that will distinguish between these two possible models.

The type of experiment we shall consider is one in which we measure

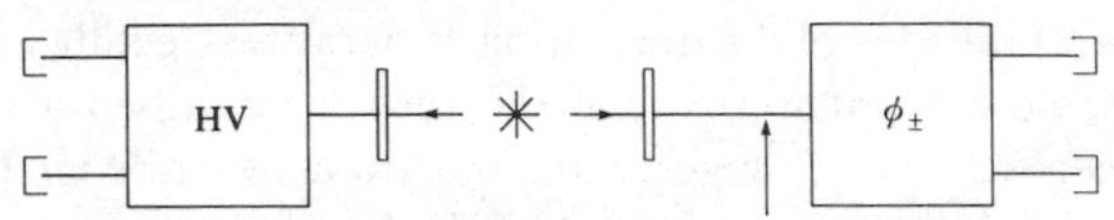

Fig. 3.3 The predictions of quantum physics for photon pairs can be tested by measuring the HV polarization of one photon and the $\phi_\pm$ polarization of the other (i.e. whether it is polarized parallel or perpendicular to a direction making an angle ϕ to the horizontal). If the left-hand measurement puts the right-hand photon into a particular HV state, the probabilities of subsequent $\phi_\pm$ measurements can be calculated.

the HV polarization of the left-hand photon, as before, but now measure the polarization direction of the second photon parallel and perpendicular to a direction at some angle ϕ to the horizontal, as in Figure 3.3. We first analyse this experiment using conventional quantum theory and suppressing any doubts we may have about action at a distance. From this point of view, when the right-hand photon reaches a point the same distance from the centre as the left-hand polarizer, its HV polarization is measured and from then on it is either horizontally or vertically polarized depending on the result of the left-hand measurement. We imagine this experiment repeated a large number (say N) times, after which we would expect about $N/2$ photons to have passed through either channel of the HV polarizer. Now it follows from our present assumptions that, corresponding to each of the $N/2$ vertically polarized left-hand photons there is a right-hand photon that is horizontally polarized before it reaches the right-hand apparatus. We can therefore calculate the number of these that would be expected to emerge in the positive channel (i.e. polarized at an angle ϕ to the horizontal) of the ϕ apparatus. We call this number $n(\nu, \phi_+)$ and, from Figure 3.4 and the general theory discussed in Chapter 2, this is just equal to $(N/2)\cos^2\phi$ while the number emerging in the other channel, $n(\nu, \phi_-)$, is just $(N/2)\sin^2\phi$. We can apply similar arguments to photon pairs whose left-hand members are horizontally polarized and we can summarize our results as

$$\left.\begin{aligned} n(\nu, \phi_+) &= \tfrac{1}{2}N\cos^2\phi, \\ n(\nu, \phi_-) &= \tfrac{1}{2}N\sin^2\phi, \\ n(h, \phi_+) &= \tfrac{1}{2}N\sin^2\phi, \\ n(h, \phi_-) &= \tfrac{1}{2}N\cos^2\phi. \end{aligned}\right\} \qquad (1)$$

We notice in passing that the total number in the positive ϕ channel – i.e. $n(\nu, \phi_+) + n(h, \phi_+)$ – is the same as the total number in the negative ϕ channel – $n(\nu, \phi_-) + n(h, \phi_-)$ – and equals $\tfrac{1}{2}N$, as we would expect because we have assumed nothing about the absolute polarization direction of the photons emitted by the source.

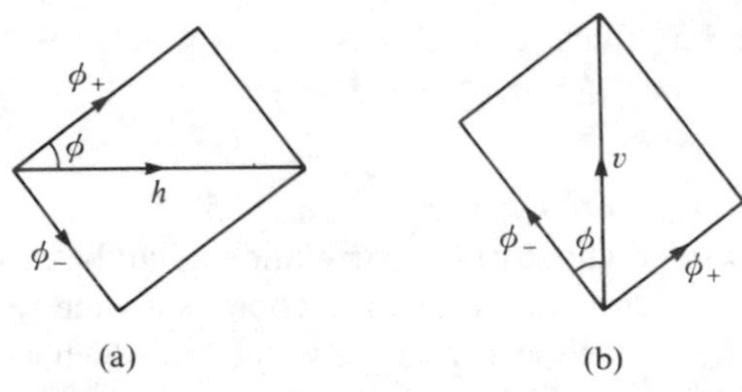

Fig. 3.4 If the left-hand photon is vertically polarized we conclude from quantum theory that the one on the right is horizontal. If the $\phi_\pm$ polarization of $\frac{1}{2}N$ such photons is measured, the number recorded as ϕ_+ will be proportional to the square of the ϕ_+ component of the electric field of a horizontally polarized wave. It follows from (a) that this number, $n(\nu, \phi_+)$, must equal $\frac{1}{2}N\cos^2\phi$ while the number emerging in the ϕ_- channel, $n(\nu, \phi_-)$, equals $\frac{1}{2}N\sin^2\phi$. Similarly it follows from (b) that $n(h, \phi_+) = \frac{1}{2}N\sin^2\phi$ and $n(h, \phi_-) = \frac{1}{2}N\cos^2\phi$.

To compare the predictions of quantum physics with those of hidden-variable theories, it is useful to calculate what is known as a 'correlation coefficient' C which is defined as

$$C = [n(\nu, \phi_+) + n(h, \phi_-) - n(\nu, \phi_-) - n(h, \phi_+)]/N. \qquad (2)$$

We can see why this is called a correlation coefficient by considering some special cases. If $\phi = 0$, both sets of apparatus are making the same measurement, in which case we know that they must always obtain the same result, and therefore

$$n(\nu, \phi_+) = n(h, \phi_-) = \tfrac{1}{2}N \text{ and } n(\nu, \phi_-) = n(h, \phi_+) = 0.$$

Using these relations and equation (2) we see that $C = 1.0$ in this case and we say the results are perfectly correlated. The case where $\phi = 90°$ is just the same except that the roles of the two right-hand channels have been reversed; it is easily seen that C is now equal to -1.0 and we have perfect 'anticorrelation'. Half-way between, when $\phi = 45°$, we would expect from symmetry that all combinations of photon pairs would be equally likely so that

$$n(\nu, \phi_+) = n(h, \phi_-) = n(\nu, \phi_-) = n(h, \phi_+) = N/4.$$

C is therefore zero in this case and we have no correlation. For general values of ϕ, C measures the extent to which the results on the two sides are correlated with each other and it is the quantum and hidden-variable predictions of this quantity that we shall shortly compare with each other and with experiment.

A quantum-physics prediction of the correlation coefficient is readily obtained using the expressions given in equations (1) and (2):

$$C(\text{quantum}) = (\tfrac{1}{2}N\cos^2\phi + \tfrac{1}{2}N\cos^2\phi - \tfrac{1}{2}N\sin^2\phi - \tfrac{1}{2}N\sin^2\phi)/N$$
$$= \cos^2\phi - \sin^2\phi$$
$$= \cos 2\phi \qquad (3)$$

A graph of this fraction is shown in Figure 3.5.

We now consider what form of correlation might be expected from a deterministic hidden-variable theory. There is a wide variety of such theories, but we shall choose a simple version in which we assume that as each photon is emitted from the atom it is plane polarized in some direction which varies at random from pair to pair and that the two photons in each pair always have perpendicular polarization. We shall also assume that a photon always emerges through the polarizer channel whose direction is nearest to the actual polarization of the photon; thus a photon approaching an HV polarizer will appear to be horizontally polarized if its actual polarization direction makes an angle of less than 45° with the horizontal and will appear vertically polarized otherwise.

We can deduce the form of the correlation coefficient expected on the basis of this model by a careful consideration of the relative polarization directions. This is illustrated in Figure 3.6 and for the moment we suppose that the left-hand polarization has been measured as vertical

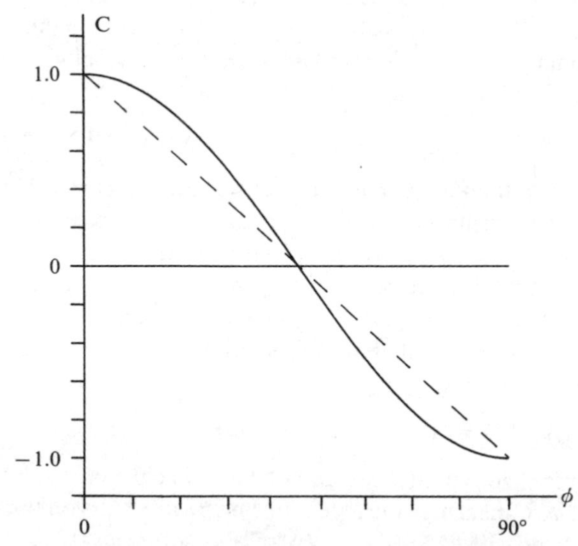

Fig. 3.5 Quantum physics predicts that the correlation coefficient $C = \cos 2\phi$ (continuous line), but a hidden-variable theory predicts that $C = 1 - \phi/45°$ (broken line). The results of experiment are in agreement with quantum theory and are inconsistent with the hidden-variable predictions.

and that on the right-hand as parallel to the direction ϕ. From the first measurement we can conclude that the polarization direction of the left-hand photon before the measurement must have been within 45° of the vertical and therefore (as the two polarizations are always at right angles to each other) the actual right-hand polarization must be within 45° of the horizontal, that is within the shaded sector of Figure 3.6(a). The second measurement implies that the right-hand photon is polarized somewhere within 45° of the direction ϕ and therefore in the shaded part of Figure 3.6(b). Putting these two conclusions together we see that the left-hand photon will be measured as vertical and the right-hand as parallel to ϕ if the actual right-hand polarization lies in the doubly shaded portion of Figure 3.6(c). As the initial polarization

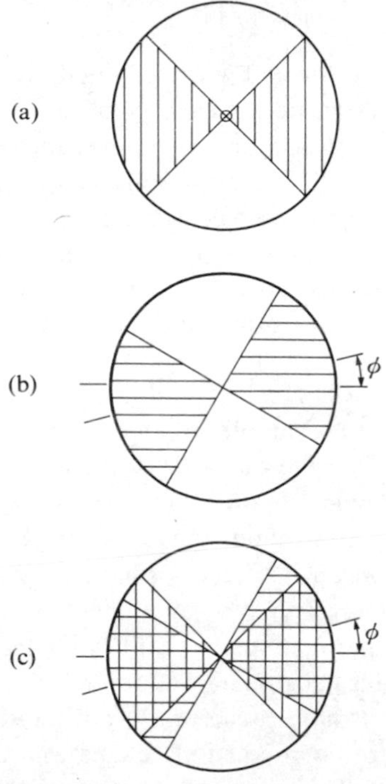

Fig. 3.6 According to the hidden-variable theory discussed in the text, if the left-hand photon is measured as vertically polarized, the polarization of the right-hand photon must lie within 45° of the horizontal – i.e. within the shaded area of (a). If a measurement of the right-hand ϕ component is positive it follows similarly that the polarization must lie within the shaded area of (b). Thus if both these results are obtained on a photon pair the right-hand polarization must lie within the doubly hatched area of (c). The fraction of the circle occupied by the doubly hatched area of (c) is clearly $(90° - \phi)/180°$.

direction varies at random from photon pair to photon pair, it follows that the number with this property is proportional to the size of this sector whose angle is clearly equal to $90° - \phi$. Similar arguments can be applied to the other possible outcomes of the measurement and collecting all results together we get

$$\left.\begin{aligned} n(\nu, \phi_+)/N &= (90° - \phi)/180°, \\ n(\nu, \phi_-)/N &= \phi/180°, \\ n(h, \phi_-)/N &= (90° - \phi)/180°, \\ n(h, \phi_+)/N &= \phi/180°, \end{aligned}\right\} \tag{4}$$

and therefore, combining equations (3) and (4),

$$\begin{aligned} C(\text{hidden variable}) &= (90° - \phi + 90° - \phi - \phi - \phi)/180° \\ &= 1 - \phi/45°. \end{aligned} \tag{5}$$

This quantity is also plotted on Figure 3.5 where we see that it agrees with the quantum prediction at the special points $\phi = 0, 90°$ and $45°$, but that there is considerable disagreement at other angles. This reaches a maximum at $\phi = 22\frac{1}{2}°$ when C(quantum) = 0.71 and C(hidden variable) = 0.50. Experiments with photon pairs are quite difficult, but a number have been performed in recent years and the results of measurements of these correlation coefficients are in good agreement with the quantum prediction and are not consistent with this hidden-variable theory.

The hidden-variable theory just described is only one of many deterministic local theories to be invented to describe the behaviour of photons. It is possible, for example, to modify the theory so that the causal connection between the outcome of the measurement and the angle between the 'real polarization' direction and the measuring axis is more complex than that considered above. We might try to effect this modification in such a way as to make the hidden-variable predictions identical to those of quantum theory – or at least so close to them as to be indistinguishable experimentally – but this would turn out to be a fruitless task. We say this because in 1969 John S. Bell showed that no hidden-variable theory which preserves locality and determinism is capable of reproducing the predictions of quantum physics for the two-photon experiment. This is a vitally important theoretical deduction that has been the mainspring of most of the theoretical and experimental research in this field for the last fifteen years. Because of this, we devote the next section to a proof of what has become known as Bell's Theorem. A reader who is uninterested in mathematical proofs and who is prepared to take such results on trust could proceed directly to the conclusions of this section which begin after equation (9) on page 41 while still keeping track of the main argument in this chapter.

Bell's Theorem

We begin by leaving the world of quantum physics for the moment to play the following game. Take a piece of paper and write down three symbols, each of which must be either a + or a −, in any order; write another three +'s and −'s under the first three and repeat the process until you have about 10 or 20 rows each containing three symbols. An example of the kind of pattern generated is given in Table 3.1.

Table 3.1

h	ϕ	θ
+	+	−
+	−	+
−	+	−
+	−	+
−	+	+
−	−	+
+	−	−
−	+	−
+	+	+
−	+	−
−	−	+
+	+	−
+	−	+

Label the three columns h, ϕ and θ as shown. The next step is to go through your list and count how many rows have a + in both column h and column ϕ and call this number $n(h = +, \phi = +)$ (in the example above it equals 3). Now count how many have a − in column ϕ at the same time as a + in column θ and call this $n(\phi = -, \theta = +)$ (= 5 in the example), and then find $n(h = +, \theta = +)$, the number of pairs with +'s in both columns h and θ (4 in the above example). Add together the first two numbers and if you have followed the instructions properly you will find that your answer is always greater than or possibly equal to the last number. That is

$$n(h = +, \phi = +) + n(\phi = -, \theta = +) \geqslant n(h = +, \theta = +). \quad (5)$$

Try again if you like and see if you can find a set of triplets composed of +'s and −'s that does not obey this relation: you will not succeed, because it is impossible.

It is quite easy to prove that relation (5) must hold for all sets of numbers of the type described. Consider first the set which has $h = +$ and $\phi = +$. This is composed of two kinds of components: those with $h = +, \phi = +$ and $\theta = +$ and those with $h = +, \phi = +$ and $\theta = -$. That is

$$n(h = +, \phi = +) = n(h = +, \phi = +, \theta = +) + n(h = +, \phi = +, \theta = -).$$

In exactly the same way

$$n(\phi = -, \theta = +) = n(h = +, \phi = -, \theta = +) + n(h = -, \phi = -, \theta = +)$$

and

$$n(h = +, \theta = +) = n(h = +, \phi = +, \theta = +) + n(h = +, \phi = -, \theta = +).$$

If we add together the first two equations we get

$$\begin{aligned} n(h = +, \phi = +) &+ n(\phi = -, \theta = +) \\ &= n(h = +, \phi = +, \theta = +) + n(h = +, \phi = +, \theta = -) \\ &\quad + n(h = +, \phi = -, \theta = +) + n(h = -, \phi = -, \theta = +) \end{aligned}$$

But the first and third terms on the right-hand side of this equation are just those which when added together make up the term $n(h = +, \theta = +)$. It therefore follows that

$$\begin{aligned} n(h = +, \phi = +) &+ n(\phi = -, \theta = +) \\ &= n(h = +, \theta = +) + n(h = +, \phi = +, \theta = -) \\ &\quad + n(h = -, \phi = -, \theta = +) \\ &\geqslant n(h = +, \theta = +) \end{aligned}$$

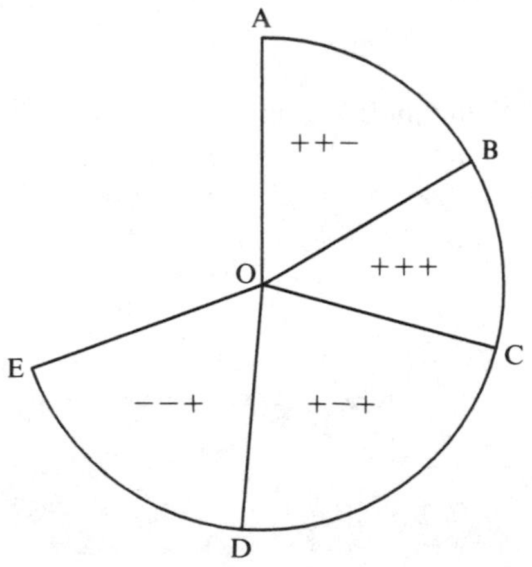

Fig. 3.7 The slices in the pie diagram represent the numbers of triplets of the types indicated. The number with $h = +$ and $\phi = +$ is therefore represented by the sector AOC. Similarly $n(\phi = -, \theta = +)$ is given by COE and $n(h = +, \theta = +)$ by BOD. Clearly AOC + COE must be greater than or equal to BOD so it follows that

$$n(h = +, \phi = +) + n(\phi = -, \theta = +) \geqslant n(h = +, \theta = +).$$

which is just equation (5). A geometrical version of the same proof is shown in Figure 3.7.

What has all this got to do with the properties of pairs of polarized photons? Imagine first that we could measure the polarization of a photon in three different directions, without disturbing it in any way, and find out whether it was: (i) parallel or perpendicular to the horizontal, (ii) parallel or perpendicular to a direction at ϕ to the horizontal and (iii) parallel or perpendicular to a third direction at θ to the horizontal. If in each case we wrote down a + when the result was parallel and a − when it was perpendicular, and if we repeated the experiment a number of times, we would get a set of numbers just like those discussed above which would therefore be subject to the relation (5). Of course we can't actually do this because we know it is not possible to make independent measurements of three polarization components of a single photon, but it turns out that all the quantities entering equation (5) can be obtained from measurements on correlated photon pairs, provided only that the hidden-variable assumptions of locality and determinism hold – i.e. provided the polarization of one photon cannot be affected by a measurement on the other a long distance away and the result of each measurement is determined by some (hidden) property of the photon.

We consider three separate experiments as set out in Figure 3.8. In the first, HV is measured on the left while $\phi_\pm$ is measured on the right. Assuming locality we can say that every time a photon is detected in the left-hand vertical channel, a photon would have been detected in the horizontal channel on the right if the right-hand apparatus had been set up this way. It follows that if this experiment is repeated a large number

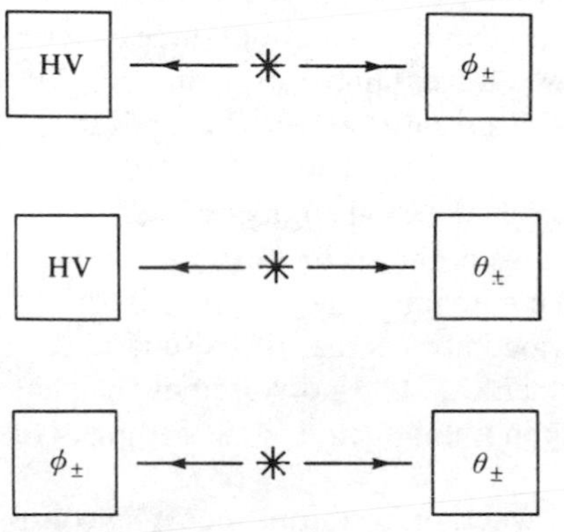

Fig. 3.8 In proving Bell's Theorem we consider three separate sets of measurements on correlated photon pairs, in which the polarizers are set in the orientations indicated.

of times a value for $n(h = +, \phi = +)$ for the right-hand photons is obtained. That is

$$n(\nu, \phi_+) = n(h = +, \phi = +). \quad (6)$$

We now consider the second experiment in which the right-hand apparatus is reoriented to make measurements in some other direction (say at θ to the horizontal) while the left-hand apparatus is left at HV. We get in exactly the same way

$$n(\nu, \theta_+) = n(h = +, \theta = +). \quad (7)$$

(Of course we have not made the two measurements on the same set of photon pairs, but provided the total number of photon pairs was large and the same for both experiments we can conclude that, if the right-hand apparatus had been in the ϕ direction for the second experiment, then the $n(\nu, \phi_+)$ obtained would have been the same as before, apart from small statistical fluctuations.) We now consider a third experiment where the *left-hand* apparatus is oriented in the direction ϕ while the right-hand apparatus is left in the direction θ. Just as before, every time we find a left-hand photon in the ϕ_+ channel we assume that the right-hand one would have been ϕ_- and we conclude that

$$n(\phi_+, \theta_+) = n(\phi = -, \theta = +). \quad (8)$$

It follows directly from these relations and from equation (5) that if our assumptions are correct

$$n(\nu, \phi_+) + n(\phi_+, \theta_+) \geqslant n(\nu, \theta_+). \quad (9)$$

This is Bell's Theorem, sometimes known as Bell's inequality. Putting it into words, it states that if we carry out three experiments to measure the polarizations of a large number of photon pairs in which the right- and left-hand polarizers are (i) vertical and at an angle ϕ to the horizontal, (ii) vertical and at an angle θ and (iii) at angles ϕ on the left and θ on the right, then the total number of pairs in which both photons are registered positive in the second experiment can never be greater than the sum of the numbers of doubly positive pairs in the other two experiments; provided always that the results of the experiments are determined by hidden variables possessed by the photons and that the state of either photon is unaffected by the setting of the other (distant) apparatus.

Let us now see whether quantum theory is consistent with Bell's inequality. The quantum expression for $n(\nu, \phi_+)$ is given in equation (1) as $\frac{1}{2}N\cos^2\phi$ and $n(\nu, \theta_+)$ is similarly $\frac{1}{2}N\cos^2\theta$. To obtain an expression for $n(\phi_+, \theta_+)$ we see that, if we rotate our axes through the angle ϕ, the direction ϕ_+ corresponds to the new 'horizontal' direction and the direction θ_+ is at an angle of $\theta - \phi$ to this. Hence $n(\phi_+, \theta_+) = n(h, (\theta - \phi)_+) = \frac{1}{2}N \sin^2(\theta - \phi)$ using (1) again. It follows that Bell's inequality and quantum physics could be consistent if and only if

$$\cos^2\phi + \sin^2(\theta - \phi) \geqslant \cos^2\theta \qquad (10)$$

for all possible values of θ and ϕ. On the other hand to prove that quantum physics is inconsistent with Bell's Theorem it is only necessary to show that equation (10) is false for some particular values of θ and ϕ. It is convenient to consider the special case where $\phi = 3\theta$ when the left-hand side becomes $\cos^2 3\theta + \sin^2 2\theta$. Figure 3.9 shows the difference between the left- and right-hand sides of (10) as a function of θ in this case, and we see that, although Bell's Theorem is obeyed for values of θ greater than 30°, it is clearly breached if θ is between 0° and 30°. In the particular case where $\theta = 20°$ and $\phi = 60°$ the left-hand and right-hand sides are 0.66 and 0.88 respectively in clear disagreement with equation (10).

We are therefore forced to the conclusion that either quantum physics does not correctly predict the results of polarization measurements on photon pairs or one of the assumptions on which Bell's Theorem is based is wrong. But these are extremely basic assumptions.

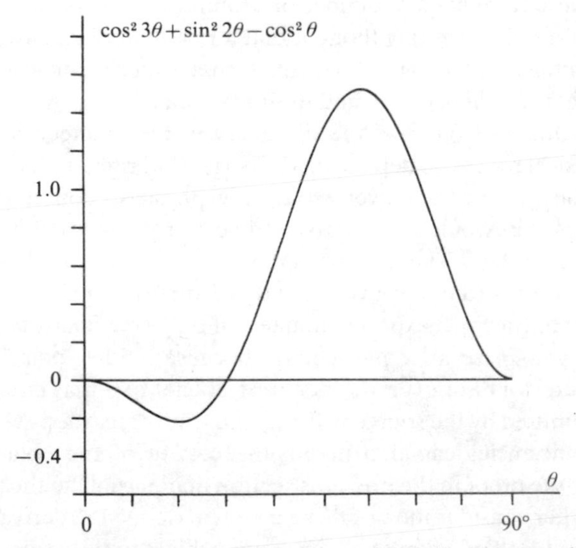

Fig. 3.9 If quantum physics were consistent with Bell's Theorem the function $\cos^2 3\theta + \sin^2 2\theta - \cos^2\theta$ would have to be positive or zero for all values of θ. The graph shows that this is not true for values of θ between 0° and 30°.

We have simply said that the outcome of a photon-measurement experiment cannot be affected by the way another distant apparatus (typically several metres away) is set up, and that the results of such a measurement are determined by some property of the photon. The only other assumptions are some basic rules of logic and mathematics! It should now be clear why experiments on the behaviour of photon pairs have been so important in recent years: they should be capable of finally determining whether physical objects are affected only by influences at the point they are at or whether the non-locality implicit in quantum physics is an inevitable fact of nature.

The experiments

It might be thought that the outstanding success of quantum theory over the whole compass of physical phenomena would have meant that the predictions of Bell's Theorem would already have been implicitly tested by experiments performed before its publication in 1969. In particular, the excellent agreement between the calculated and observed properties of helium, whose atoms each contain two electrons, might be thought to be sensitive to the properties of correlated pairs of particles. However, it soon became apparent that no direct test of the particular kind of correlation involved in the Bell inequality had actually been made. There have been a number of examples in the development of physics of it being wrongly thought that a possibility had already been experimentally tested, and when an actual experiment was done it showed that the theory accepted until then was wrong. A well known recent example of this was the discovery in the nineteen-fifties that some physical processes depend on the parity (i.e. right- or left-handedness) of the system. Thus, even though few physicists doubted whether quantum physics would turn out to be correct, it was very important that a direct test of Bell's Theorem be made.

It was soon realized, however, that there are severe practical difficulties preventing a direct experimental test of Bell's inequality in the form given above. These arise particularly because neither polarizers nor photon detectors are ever 100 per cent efficient so that many of the photons emitted by the source will not actually be recorded. Moreover, these inefficiencies can depend on the setting of the polarizers so rendering the proof in the previous section inapplicable to the practical case. Further consideration of these problems led to the derivation of a new form of Bell's Theorem which is not subject to this criticism. This involves consideration of an experiment in which measurements are made with four relative orientations of the polarizers instead of the three settings considered above. We shall not discuss the proof of this extension of Bell's Theorem, but simply state the result that

$$n(\nu, \phi_+) - n(\nu, \psi_+) + n(\theta_+, \phi_+) + n(\theta_+, \psi_+) \leqslant n(\theta_+) + n(\phi_+) \tag{11}$$

where the left-hand polarizer can be set to measure either HV polarization or polarization at an angle θ to the horizontal and the two orientations of the right-hand polarizer are ϕ and ψ. The numbers on the left-hand side of equation (11) refer to the number of times photons are simultaneously recorded in the appropriate channels while the numbers on the right-hand side of equation (11) refer to measurements made with one of the polarizers removed: thus $n(\theta_+)$ represents the number of times a photon appears in the left-hand θ_+ channel at the same time as a right-hand photon is recorded with no polarizer present and $n(\phi_+)$ is defined similarly. Clearly six separate experiments are required to collect the data relevant to equation (11) and it is assumed that each one lasts for the same length of time so that the same total number of photon pairs is involved in each case.

The extended Bell inequality has a further important property which we shall also state without proof. This is that it tests not only deterministic theories, but also a wide range of hidden-variable theories that include an element of randomness, such as the guiding-wave model referred to near the end of Chapter 2. If the relation stated in equation (11) is breached, any kind of hidden-variable theory that preserves locality is ruled out.

A number of experiments testing Bell's Theorem in this or similar ways have been carried out over the past fifteen years or so. Although one of the early experiments initially produced results consistent with Bell's Theorem and in disagreement with quantum theory, all the others (including a repeat of the experiment just referred to) agree with quantum predictions and breach the Bell inequality. The experiments have been performed by a number of workers, but the most recent, and probably the most definitive, have been performed by Alain Aspect in France. In 1982 he reported an experiment just like that described above in which the angles were

$$\theta = 45°; \qquad \phi = 67\tfrac{1}{2}°; \qquad \psi = 22\tfrac{1}{2}°$$

When the results of these experiments are substituted into equation (11) it is found that the left-hand side, instead of being smaller than the right-hand side, is actually larger by an amount equal to $0.101N$, where N is the total number of photon pairs in each run. A quantum calculation of the same quantity (taking detection efficiency into account) produces

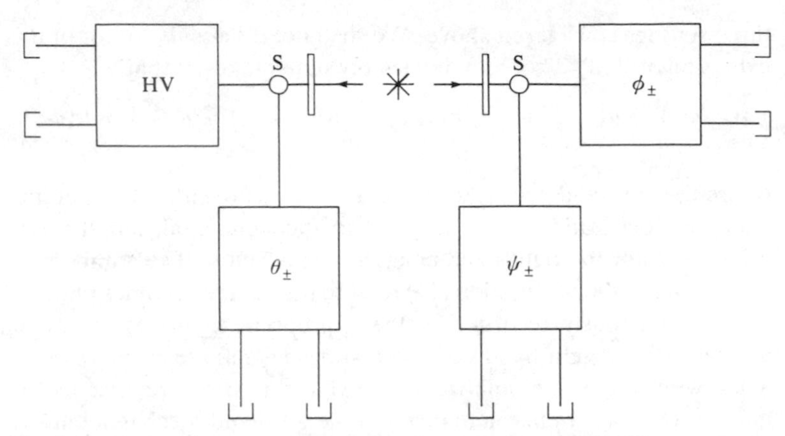

Fig. 3.10 In the Aspect experiment to test the predictions of Bell's Theorem for the polarization of photon pairs, either the HV or the $\theta_\pm$ polarization can be measured on the left and either $\phi_\pm$ or $\psi_\pm$ on the right. The ultrasonic switches, S, operate so quickly that it is impossible for the photons to be influenced by the settings of the distant apparatus unless such influences can be propagated faster than light. The results of this experiment agree with the predictions of quantum physics and are inconsistent with any local hidden-variable theory.

the result $0.112N$. The estimated errors in the experiment are large enough to embrace the quantum result, but are small enough to rule out the Bell inequality.

The Aspect experiment possesses an additional feature of considerable interest. This is illustrated in Figure 3.10 from which we see that the two polarization measurements on each side were actually made using different polarizers which were always in position, the photons being switched from one polarizer to the other by the devices marked S in order to perform the appropriate measurements. These 'switches' are actually actuated by high-frequency ultrasonic waves so that the measurement is switched from one channel to another about one hundred million times per second! This rapid switching has an important consequence. Remember that at the beginning of our discussion we showed how the quantum result follows from the fact that the right-hand photon is apparently influenced by the setting of the left-hand apparatus. We might envisage, therefore, that the left-hand apparatus is sending some kind of 'message' to the right-hand photon telling it how it is set up so that it can interact in an appropriate way with the right-hand polarizer. The switching experiment shows that any such message must travel in a time less than one hundred millionth of a second – otherwise the right-hand photon would behave as if it had been measured by a polarizer set in an earlier orientation of the left-hand apparatus. But in Aspect's experiment, the two polarizers are about 10 metres apart and a signal travelling at the speed of light would take about three times the switching time to cover this distance. Now it is well known that no physical object or signal can travel faster than light so the possibility that the photon is receiving messages from the other detector must also be discounted.

Considerable thought has gone into a critical appraisal of the design details of experiments such as those discussed to see if the results are really inconsistent with Bell's Theorem or if some loophole in the reasoning remains. The only conceivable possibility seems to be that the efficiencies of the photon detectors might somehow depend on the hidden variables. This would not of course be possible if the detectors always registered every photon entering them, but real detector efficiencies are quite low so this loophole cannot be said to be completely closed. However, the improbability of this hypothesis, combined with the fact that the experimental results are not only in breach of Bell's Theorem but also in excellent agreement with quantum theory, has convinced nearly everyone working in this area that all local hidden-variable theories can now be discounted.

Discussion

It should now be clear why the results of two-photon experiments lead some people to describe the Einstein–Podolski–Rosen effect as a 'paradox'. Although some scientists strongly disagree with this label, there is something very paradoxical about the position we have got ourselves into. We started off by showing that quantum physics implies that an operation carried out on one photon affects the state of another a long distance away. We have considered the alternative possibility that the effect arises from some property possessed by each photon when the pair is created, but Bell's Theorem and the associated experiments have forced us to reject this. Now the final Aspect experiment proves that no message can be passed to a photon from the distant apparatus. At first sight this seems to contradict the initial statement that we thought had just been confirmed!

There is certainly no easy answer to the problem of non-locality and the EPR paradox. In the next chapter we shall return to this question in the context of a deeper discussion of the conventional or 'Copenhagen' interpretation of quantum theory, but for the moment we make several important points that must be kept in mind in any discussion of this question.

First, although the photon certainly appears to be influenced by the distant apparatus, this is not the sort of influence we are used to encountering between physical objects. In particular it is not the sort of influence we can use to transmit information or signals from one place to another. We can see quite simply why this is so by considering two experimenters at opposite ends of a two-photon apparatus trying to use the equipment to pass signals to each other. The point is that, however the left-hand apparatus is oriented, the photons emerge through the two channels of the right-hand polarizer at random. This follows from the fact that even if the left-hand polarizer (assumed to be HV for the moment) puts one right-hand photon into a state of vertical polarization, it will put the next into the same or the opposite state *at random*; thus the photons emerge at random from the two channels of the right-hand polarizer whatever its orientation. We can therefore conclude that there is no way an experimenter can draw any conclusions about the orientation of the other apparatus by observing only his 'own' photons. It is only when we bring together the results of the measurements at both ends and examine correlations that the effects discussed in this chapter are observed.

Secondly, the fact that it is impossible to use the EPR set-up to transmit information, at least partly resolves the problem of influences apparently travelling faster than light. If these influences cannot be used to transmit information, then they need not be subject to the theory of relativity which requires that no *signal* can be transmitted faster than light. We are dealing with a *correlation* between two sets of events which does not travel in one direction or the other. When we first analysed the situation in which the left-hand apparatus measured HV polarization while that on the right made measurements at an angle ϕ to the horizontal (see Figure 3.3 and equations (1) and (2)) we said that the left-hand apparatus had put the right-hand photon into an H or V state that was then further analysed by the right-hand polarizer. But if we had put it the other way round and considered the right-hand measurement first, we would have obtained the same answers for all the experimental results such as $n(\nu, \phi_+)$ etc. and the correlation coefficients. Indeed readers familiar with the theory of relativity will realize that an observer travelling past the apparatus in Figure 3.3 from left to right at a sufficiently high speed would conclude that the right-hand photon had been detected before that on the left and therefore that the ϕ measurement had occurred first. As the measured correlations do not depend on any assumed direction of travel, there is no inconsistency between the EPR experiments and the theory of relativity.

Finally we note that, although the Aspect experiments confirm the predictions of quantum physics, their significance is even more far reaching than this. The fact that Bell's inequality is breached means that no theory that preserves locality can ever be consistent with experiment. Even if quantum theory were shown to be incorrect tomorrow, any new fundamental theory would also have to face the challenge of the violation of Bell's inequality and would have to predict the observed correlations between widely separated measurements

The aim of this chapter has been to demonstrate that it is impossible to avoid the revolutionary conceptual ideas of quantum physics by postulating any kind of hidden-variable theory that preserves locality: the observed properties of pairs of photons cannot be explained without postulating some correlation between the state of a measuring apparatus and that of a distant photon. However, what is now the orthodox approach to quantum physics takes an even more radical point of view than this, questioning whether such a postulate is meaningful and if the photons can be said to have any existence at all until they are observed. This viewpoint is known as the Copenhagen interpretation and is the subject of the next chapter.

4 · Wonderful Copenhagen?

The 1935 paper by Einstein, Podolski and Rosen represented the culmination of a long debate that had begun soon after the quantum theory was developed in the nineteen-twenties. One of the main protagonists in this discussion was Niels Bohr, a Danish physicist who worked in Copenhagen until, like so many other European scientists of his time, he became a refugee in the face of the German invasion during the Second World War. As we shall see, Bohr's views differed strongly from those of Einstein and his co-workers on a number of fundamental issues, but it was his approach to the fundamental problems of quantum physics that eventually gained general, though not universal, acceptance. Because much of Bohr's work was done in that city, his ideas and those developed from them have become known as 'The Copenhagen Interpretation'. In this chapter we shall discuss the main ideas of this approach and try to appreciate their strengths as well as trying to understand why it still seems to leave unanswered some important questions that will form the subject of the later chapters of this book.

When Einstein said that 'God does not play dice', Bohr is said to have replied 'Don't tell God what to do!' The historical accuracy of this exchange may be in doubt, but it encapsulates the differences of approach of the two men. Whereas Einstein approached quantum physics with doubts and sought to reveal its incompleteness by demonstrating its inconsistency, Bohr's approach was to accept the quantum ideas completely and to explore their consequences for *our* ways of thinking about the physical universe. Central to the Copenhagen interpretation is a distinction between the microscopic quantum world and the everyday *macroscopic* apparatus we use to make measurements. The only information we can have about the quantum world is obtained from these measurements which always have an effect on the system being measured. It is therefore pointless to ascribe properties to an isolated quantum system, as we can never know what these are: real physical properties are possessed only by the combined system of microscopic object *plus* measuring apparatus.

We can demonstrate these ideas more clearly by again considering the process of the measurement of the polarization of a photon. Suppose, as in Chapter 2, a photon approaches an HV apparatus and emerges in one of the channels – say the V channel. According to Bohr it is incorrect to speculate on what the photon polarization was before the measurement as this is unknowable and any attempt to measure it will interfere with it. After the measurement, on the other hand, it is meaningful to say that the photon is vertically polarized because if we pass it through a second HV apparatus we know that it will certainly emerge through the V channel. If, however, we direct the vertically polarized photon towards a polarizer set at some other angle – say at 45° to the horizontal – then, until we have made this second measurement we do not know through which ±45° channel the photon will emerge and it is incorrect to attribute any reality to the idea of the photon possessing 45° polarization in advance. Moreover, after the 45° measurement has been made any knowledge of the HV state will have been destroyed and it is then incorrect to attribute *this* property to the photon.

The fact that a measurement generally destroys all knowledge of some other property of a quantum system was described by Bohr as 'complementarity'. Thus HV and ±45° polarization are referred to as complementary variables whose values could never be simultaneously measured and therefore should never be simultaneously ascribed to a photon. Of course, as was indicated in Chapter 2, this is rather unsurprising as the idea of a classical wave being simultaneously vertically polarized and polarized at 45° to the horizontal is a contradiction in terms. But complementarity goes a little bit further than this. If it is wrong to ascribe a particular 45° polarization to a vertically polarized photon, it follows that no hidden variable exists to determine the result of the 45° measurement. Bohr willingly embraces this fundamental indeterminism of quantum physics and, rather than trying to recover a mechanistic model through some form of hidden-variable theory, he treats complementarity and indeterminism as fundamental facts of nature which our studies of subatomic phenomena have led us to appreciate.

The wave analogy may make the complementary nature of different photon polarization directions appear reasonably acceptable, but the application of the idea to other physical systems requires a much more radical change in our thinking. Thus in Chapter 1 we showed how it is impossible to make simultaneous precise measurements of the position and momentum of an electron: measurement of one quantity inevitably renders unpredictable the result of a subsequent measurement of the other. The Copenhagen interpretation says in this case that it is not meaningful to think of the electron as 'really' possessing a particular position or momentum unless these have been measured; and if its momentum, say, has been measured it is then meaningless to say that it is in any particular place. So also with wave–particle duality (see Chapter 1, page 9): when light or an electron beam passes through a two-slit apparatus it behaves as a wave because in these circumstances it *is* a wave; when, however, it is detected by a photographic plate or a counter it behaves like a stream of particles because in this context, when interacting with this apparatus it *is* a stream of particles. The possible outcomes of a measurement are determined by the object and the measuring apparatus together; we must not ascribe properties to the object alone unless these have been measured.

We might wonder how we know that a quantum object exists at all in the absence of any measurement. The answer is that we don't. Until we have measured some property of a system it is meaningless to talk about its existence. When some property has been measured, however, it is meaningful to talk about the existence of the object with this property until some complementary property has been measured. Thus it is usually meaningful to attribute a definite mass and charge to an electron whose existence is thereby established, and most subsequent measurements will leave these properties unaltered. In some circumstances, however, such as when an electron collides with and annihilates a positron to produce two γ-ray photons, even these quantities change in a quantum manner and lose their significance.

Copenhagen and EPR

The differences between Bohr and Einstein in their approach to what was then the still new subject of quantum physics led to a lively debate between the two that was conducted at several scientific conferences and in the scientific literature of the time. On many occasions Einstein would suggest a subtle experiment by which it seemed that the values of a pair of complementary variables could be simultaneously measured, and Bohr would reply with a more careful analysis of the problem, showing the simultaneous measurement to be impossible. In 1935, however, came the paper by Einstein, Podolski and Rosen which we discussed at length in the previous chapter. This showed how quantum physics requires that a property, such as the polarization of a photon, could be measured at a distance by measuring the polarization of a second photon that had interacted with the first some time previously. If it is inconceivable that this measurement could have interfered with the distant object, it follows that the first photon must have possessed the measured property *before* the measurement was carried out. As the property measured can be varied by the experimenter adjusting the distant apparatus, EPR concluded that all physical properties (in our example values of polarizations in all possible directions) must be 'real' *before* they are measured, in direct contradiction to the Copenhagen interpretation.

According to Bohr's colleague Leon Rosenfeld writing in 1967, the EPR paper was an 'onslaught that came down upon us like a bolt from the blue'. Bohr immediately abandoned all other work to concentrate on refuting the new challenge. Eventually he succeeded (to his own satisfaction at least) and commented to Rosenfeld 'They [EPR] do it smartly, but what counts is to do it right'.

How then did Bohr 'do it right'? How is the Copenhagen interpretation able to resolve the paradoxes described in the last chapter, problems that bothered the minds of physicists thirty years after Bohr, leading to the Bell Theorem and the Aspect experiments? The key point of Bohr's reply is that in this example the quantum system consists of two photons which must not be considered as separate entities until after a measurement has been made to separate them. It is therefore wrong to say that they have not been disturbed by the left-hand measurement as it is this that first causes the separation to occur. He also points out that the indirect method of measurement causes no breach in the rules of complementarity because, having chosen to measure (say) the HV polarization on the left, it is this component only whose value we have determined on the right-hand photon. We can get a value for a different right-hand component only by performing a further measurement that will cause a further disturbance to the system. Since the result of this further measurement cannot be predicted in advance it is incorrect of EPR to deduce that it is 'real' before the final measurement.

We see here the central idea of the Copenhagen interpretation: a quantity can be considered real only if it has been measured or if it is in a measurement situation where the outcome of the experiment is predictable. It follows that the real properties of a quantum system can be changed by an experimenter who rearranges her apparatus. As Bohr puts it 'there is essentially *the question of an influence on the very conditions that define the possible types of prediction regarding the future behaviour of the system*' (Bohr's italics).

It is well worth pausing at this point to consider the implications of this statement of Bohr's because they go to the very heart of quantum measurement theory. There are effectively three different levels of operation in a quantum measurement. The first consists of the way the measuring apparatus is set up (e.g. which components of polarization are being measured by the Aspect polarizers). The second level is the statistical result that is obtained after a large number of measurements have been made (e.g. the correlation coefficients), and the third is the result actually obtained in a particular, individual measurement. As far as this last is concerned (apart from special circumstances such as when a previously known polarization component is remeasured) this is completely random and unpredictable. At the second level, quantum physics allows us to predict future statistical behaviour if the present state is known: thus if a large number of 45° photons pass through an HV apparatus we know that half of them will appear in each channel. At the first level, the way the apparatus is set up determines what type of property will be measured and therefore, as Bohr says, what 'possible types of prediction regarding the future behaviour of the system' can be made. In the Aspect experiment it is this first level which is changed one hundred million times per second by the ultrasonic switch. As far as a measurement on an individual photon pair is concerned the results of the third-level process are random and unpredictable whatever the setting of the apparatus. The second-level statistical predictions are affected by the first-level changes in a way that can be predicted by quantum theory and Bohr would certainly not have been surprised that these predictions are confirmed by experiment.

With the benefit of the further insights from the work of Bohm, Bell and Aspect discussed in the last chapter, does the Copenhagen interpretation of EPR still hold up? From one point of view, certainly. The experimental results are entirely consistent with the quantum predictions and the 'possible types of prediction' are indeed influenced by the experimental conditions – even when these are altered one hundred million times a second in the Aspect experiment. Yet rereading Bohr's paper it does seem that he has missed or avoided the crucial point. Yes, the measurement does affect the system, but it affects the *whole* system including the distant photon. We have seen that non-locality is an essential feature of any model giving results in breach of Bell's Theorem so that some form of 'action at a distance' is necessary. EPR assume this is impossible and deduce that the quantum model must be wrong. Bohr assumes that the quantum model is correct and therefore by implication that instantaneous correlations between distant parts of a quantum system do occur. We have to accept the non-locality of such correlations as an intrinsic fact of nature revealed by the quantum process. Indeed even if we go a little further in the spirit of the Copenhagen interpretation and say that in the same way as it is meaningless to think of a single photon as having a particular polarization until it is measured it is also wrong to think of the two photons as having any independent existence until a measurement has been made, the problem of non-locality remains. Some of the difficulties arise only when we try to extrapolate beyond the actual measured reality and to attribute 'reality' to the photons before they interact with the apparatus. The Copenhagen interpretation prohibits this and considers any such unmeasured properties to be unreal and meaningless, but all conceptual problems are not then automatically resolved.

We see that the Copenhagen interpretation involves a complete revolution in our thought compared with the classical approach and it is this psychological change that Bohr believed is forced on us by the development of quantum physics. Indeed, as was mentioned earlier, most modern undergraduate courses in physics seem to be aimed at conditioning students to think in this unfamiliar way. Most of us adapt to this quite well, but some are never convinced. Einstein himself reacted to Bohr's reply with the comment that Bohr's position was logically possible, but 'so very contrary to my scientific instinct that I cannot forego my search for a more complete conception'. So far no such 'more complete conception' has been found and we seem to have to make the best of the Copenhagen interpretation. It turns out, however, that this leads us to another major problem whose conceptual and philosophical implications far exceed anything discussed so far. It is this 'measurement problem' that we shall outline in the remainder of this chapter and discuss in the rest of this book.

The measurement problem

To understand the nature of the measurement problem in quantum physics we return again to an example of polarization measurement that we discussed briefly near the end of Chapter 2 (page 24) and, since it is so important, we set it out again here. We consider a photon whose polarization is known from a previous measurement to be at 45° to the horizontal passing through an HV polarizer such as calcite crystal. The question we ask is 'Does the calcite crystal actually measure the photon polarization?' The obvious reply must be 'yes', and if we ask 'How do we know?' the answer surely is that further measurements of the

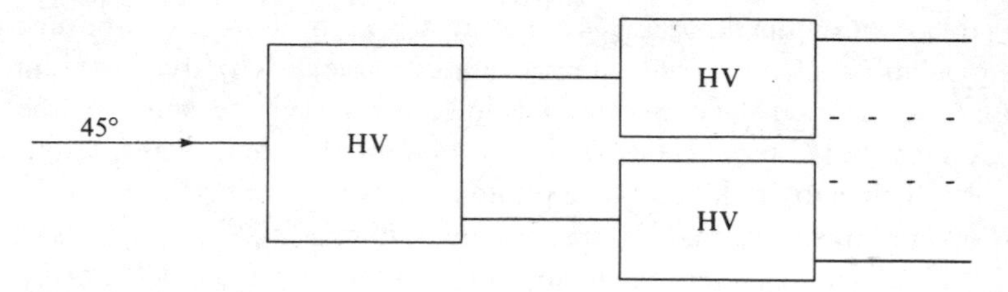

Fig. 4.1 If photons known to be polarized at 45° to the horizontal are passed through an HV polarizer they emerge at random through the horizontal and vertical channels. After this the photons are apparently either horizontally or vertically polarized, as is confirmed by further HV measurements.

properties of the photons emerging in the two channels will confirm them to be horizontally and vertically polarized respectively. In particular, if the photons are passed through further HV polarizers, all those emerging from the first one in the horizontal channel pass through the same channel of the others (Figure 4.1).

Now, however, consider the experiment described at the end of Chapter 2 and illustrated again in Figure 4.2. A beam of photons polarized at 45° to the horizontal is passed through a polarizer oriented to measure HV polarization as before, but the photons emerging from the two polarization channels are brought into the same path by a reversed calcite crystal so that when the final beam is examined it is impossible to tell through which channel a particular photon passed. If the HV apparatus has indeed made the measurement we expect, the emergent beam will be a mixture of horizontally and vertically polarized photons, and if this is now passed through a further ±45° polarizer we may expect these to emerge at random in the +45° and −45° channels. In reality, however, this does not happen: provided the apparatus is carefully set up so that the distances travelled along the two paths through the HV crystals are exactly the same (and it should be emphasized that this is a perfectly practicable experiment that has been

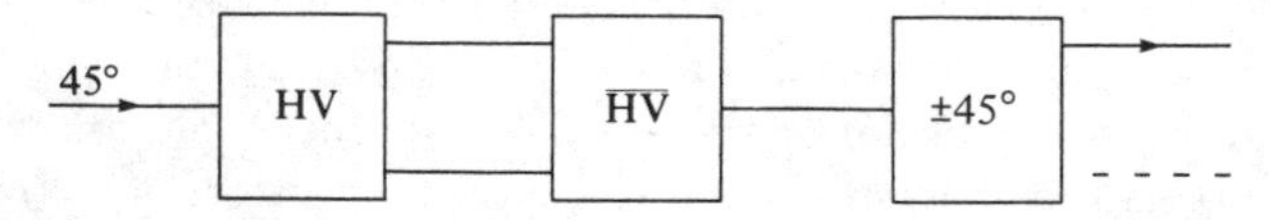

Fig. 4.2 If the photons emerging from the HV polarizer are passed through a reversed polarizer the original 45° state can be reconstructed. We must conclude that a quantum measurement of polarization cannot be performed using an HV polarizer alone.

performed many times) the photons emerging from the second crystal are all found to be polarized in the +45° direction, just as they were before entering the HV apparatus. The effect of the central 'measurement' has been entirely cancelled out! We do not know through which HV channel a particular photon passed, so we have not made a measurement of the HV polarization at all.

It can easily be seen that an effect exactly like that described above is what might be expected on the basis of the wave theory of light. Thus from this point of view the HV polarizer takes a 45° polarized light wave and splits it up into two components. The amplitude of these components oscillates up and down (or from side to side) but when they are reunited after travelling the same distance, they are back in step and recombine to form a beam whose polarization is identical to that of the original one (Figure 4.3). From the photon point of view, however, the explanation is much less clear. We know from experiments with photon counters that it is impossible to split a photon into two: it goes through either the horizontal or the vertical channel, but not both. Moreover, we cannot attribute the reconstruction of the original polarization to some interaction between photons that have travelled different paths, because exactly the same result is obtained if the light used is so weak that only one photon could be in the apparatus at any one time.

We are forced to the conclusion that a polarization apparatus such as a calcite crystal is not sufficient to make a measurement of the polarization of a photon. Besides the beam splitter we need a device to detect and record which path the photon has passed along. There are a number

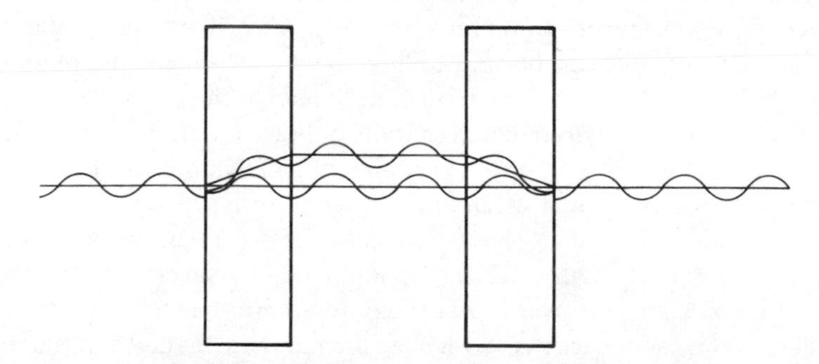

Fig. 4.3 From the wave point of view the reconstruction of the 45° state is easily understood. The first calcite crystal divides the initial 45° wave into horizontal and vertical components (both of which are drawn in the plane of the paper in the interest of clarity). These two are brought together by the second crystal: they emerge in step and recombine to re-form a 45° polarized wave.

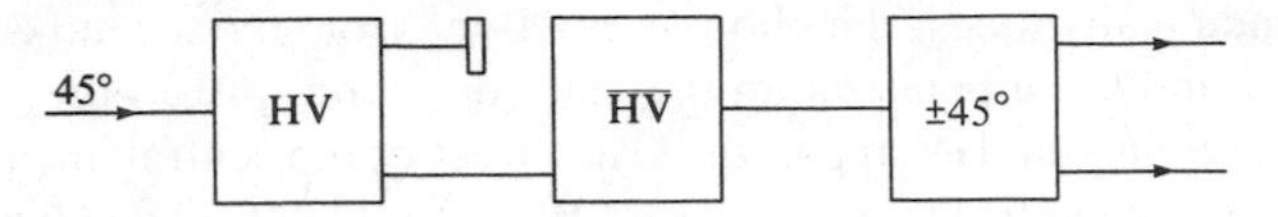

Fig. 4.4 If one of the paths between the HV and $\overline{\text{HV}}$ apparatuses is blocked, the 45° polarization can no longer be reconstructed and the photons emerge at random through the two channels of the 45° polarizer.

of ways of doing this and perhaps the most direct is just to block off one of the two paths – say the vertical one – with a shutter. Sure enough, all the photons emerge in the horizontal channel and pass randomly through either channel of the subsequent ±45° polarizer (Figure 4.4). A similar thing happens if we block off the horizontal channel, and if we were to set up some mechanical system to move the shutter in and out of each path in turn we would find that the emerging beam would indeed consist of a mixture of horizontally and vertically polarized photons. We might now try something more subtle and, instead of using a shutter, put some kind of photon detector in the beam that still allows the photon through: we might then expect to be able to record the HV polarization without destroying the ±45° polarization. However, it turns out that this is impossible: any such detector always affects the photons in such a way as to destroy the original ±45° polarization, and they emerge at random through both channels of the 45° analyser.

Let us now look at this example from the point of view of the Copenhagen interpretation. We saw earlier how Bohr always stressed the importance of measurement and warned against ascribing reality to unmeasured properties of quantum systems. Clearly before a polarization measurement can be said to have been performed, the photon must have actually been *detected* in one channel or the other. That is, a photon can be considered to be vertically or horizontally polarized if its passage through the appropriate channel of an HV polarizer has been recorded on a detector. But, in the absence of such a record, either we do not know what its polarization is or, if we knew what it was previously (as in Figure 4.2), we should not be surprised that the insertion of a polarizer *without* a detector does not change the previous polarization. Indeed we may wish to go further than this and say that, in the absence of a measurement, it is meaningless to think about the photon emerging from this polarizer as having any polarization. Only when the photon has passed through a polarizing apparatus *which includes a detector* should we consider the concept of photon polarization to be meaningful at all.

The above arguments summarize the conventional view of quantum measurement theory. If applied consistently, the correct answers are obtained in all practical situations and for many physicists that is the end of the story. But this approach is subject to a major objection, the nature of which will be discussed in the rest of this chapter and the implications of which are the subject of the rest of this book. The essential problem arises because quantum physics is the most fundamental theory we know and, if it is completely fundamental, it should be universally applicable. In particular, quantum physics should be able to explain the properties not only of atomic-scale particles such as photons, but also of macroscopic objects such as billiard balls or motor cars or photon detectors. Of course we were able to discuss the physics of such large-scale objects long before quantum ideas were ever developed: Newtonian mechanics and Maxwell's electromagnetism are normally all that are required. But it turns out that in such cases the results of quantum analysis are just the same as those obtained in conventional ways. Thus, although it would often be an unnecessarily elaborate way of obtaining the same answer, we could in principle use quantum rather than classical theory to analyse any physical situation. We seem able to conclude, therefore, that quantum physics is the final fundamental theory describing the behaviour of the physical universe: whenever its results are calculable they are always found to be in agreement with experiment, be it in the explicitly quantum regime of subatomic particles or in the macroscopic world of everyday objects where quantum and classical predictions are the same. Despite all this, however, further consideration of the measurement situation will soon show that it is impossible to apply pure quantum theory in a consistent way to all these situations and that, if there is to be such a fundamental universal theory, it is not quantum physics.

To understand this crucial point we consider again the polarization measuring experiment now modified to include a detector as in Figure 4.5. The detector is imagined to be connected to a meter with a pointer that can be in one of three possible positions: position O corresponds to the initial state before a photon has passed into the apparatus, while positions H and V correspond to a photon having passed through the horizontal and vertical channels respectively. At least this is how the apparatus behaves if we consider it as a measuring instrument. Now consider the whole assembly of counter and photon together as a single system on which we make quantum measurements. There are two possible states after the photon has passed through: the first corresponds to a horizontally polarized photon and the pointer at

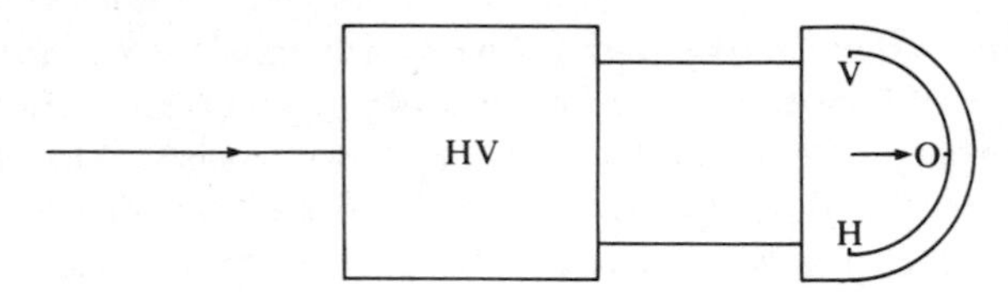

Fig. 4.5 We imagine an HV apparatus connected to a detector and meter arranged so that the pointer moves from the position O to V if the photon is detected as vertically polarized and to the position H if it is found to be horizontal. However, quantum physics implies that, in the same way as the photon is neither horizontally or vertically polarized until it has been measured, the pointer is at neither H nor V until a measurement has been made on it.

position H while in the second the photon is vertically polarized and the pointer is at V. But the same argument which told us that the photon could not be said to be vertically or horizontally polarized until we recorded through which channel it had passed can now be applied to the whole system: the counter cannot be said to be in position H or position V *until this position is measured*. Unless this measurement is made, it is always possible to envisage a mechanism similar in effect to the reversed polarizer in Figure 4.2. This would recombine the two beams into the original 45° state – as in Figure 4.2 – and restore the pointer to the position O. Only if we make a measurement of the pointer position (e.g. by placing a camera near the apparatus so that it takes a photograph of the pointer before its state is restored) is this possibility removed. But this can only be a temporary solution, because the camera can also presumably be treated as a quantum object whose state is known only when a measurement is made *on it*. This argument can be continued indefinitely and there seems to be no unique point at which the measurement can actually be said to have occurred.

The central point of the quantum measurement problem can be summarized as follows. Our analysis of the behaviour of microscopic objects like photons shows us that contradictions must arise if we attribute properties (such as polarization) to them unless these have been measured. But if quantum physics is to be a universal theory, it must be applicable to the measuring apparatus also, which therefore must not be said to be in any particular state until a measurement has been made on it.

Another way to put the problem is that a 45° photon which has passed through an HV apparatus without a detector has the potential to behave as if it were horizontally or vertically polarized, or, if it is passed through a reconstructing apparatus, as if it still had 45° polarization. Only after a measurement has been made is one of these potentialities destroyed. In exactly the same way, if quantum physics is a universal theory the detection apparatus must retain the potential of being in either of the two pointer positions until a measurement is made on it. The practical problems involved in demonstrating the reconstruction of a counter state, analagous to the reforming of the 45° photon state, are immense: real measuring apparatus and pointers are constructed from huge numbers of atoms and, before such an effect could be demonstrated, these would all have to be returned to precisely the same condition they were in before the photon entered the apparatus, which is quite impossible in practice. Unless, however, it is for some reason impossible *in principle* there is no point at which we can say that the measurement has been made.

Schrödinger's cat

The problems arising when we consider the effects of measurement on a quantum system were graphically illustrated by Erwin Schrödinger who was one of the founders of quantum mechanics. He imagined a situation similar to that set out in Figure 4.6. Inside a large box we have, as well as the familiar light source, polarizer and detector, a loaded revolver (or some other lethal device) and a cat! Moreover, the pointer on the detector is now connected to the trigger of the loaded revolver in such a way that if a vertically polarized photon is detected the revolver fires and the cat is killed, whereas a horizontally polarized photon does not affect the revolver and the cat remains alive. When shut, the box containing the cat and apparatus is assumed to be perfectly opaque to light, sound or any other signal that could tell us what is happening inside. We now ask what will happen when a single photon is emitted by the light source. If we regard the cat as a measuring apparatus the answer is straightforward: the cat is killed if the photon is vertically polarized and remains alive if the polarization is horizontal. But what does an observer outside the box who accepts the Copenhagen interpretation say? Presumably he cannot draw any conclusion about the state of the system until it has been measured, which, as far as he is concerned, is when the box has been opened and the state of the cat (dead or alive!) observed. More than this, he will conclude that, until this observation has been made, a further operation is always in principle possible to restore the photon and the box contents to their original condition and therefore we cannot say that the state of the

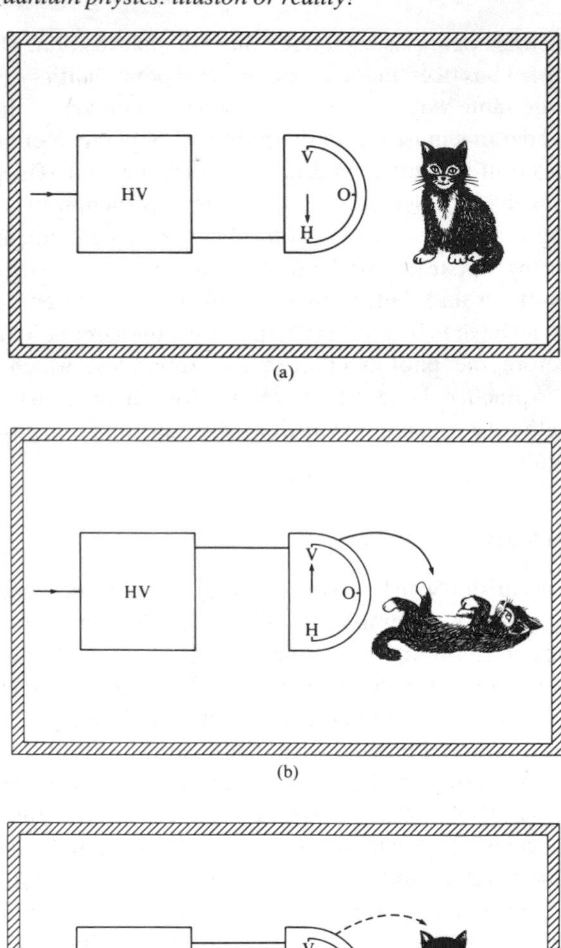

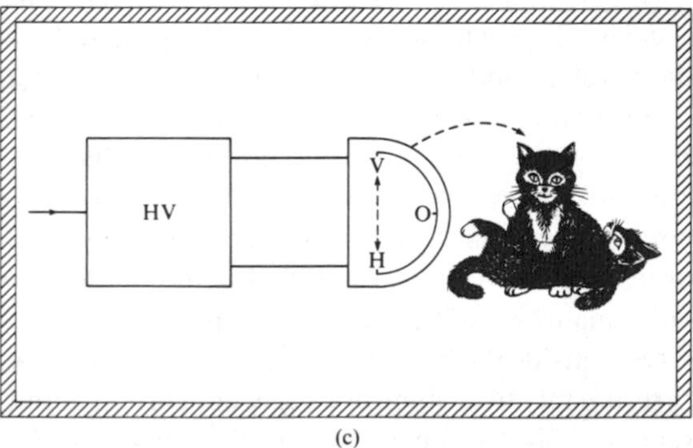

Fig. 4.6 Schrödinger's cat. If a photon comes out through the horizontal channel of the polarizer the cat is unaffected and remains alive (a), but if it is vertical a lethal device is triggered and the cat is killed (b). Does quantum physics imply that until the box is opened and its state measured, the cat is neither alive nor dead?

system has been changed: the photon is still polarized at 45° to the horizontal and the cat is presumably in a state of suspended animation until it is observed!

We can illustrate the point in an even more dramatic way by considering another example of a quantum measurement. It is well known that the evolution of living organisms results from mutation in the DNA of the genetic material of members of a species, which in turn causes a change in the characteristics of the offspring. It is also a fact that such mutations can be caused by the passage of high-energy cosmic ray particles. But these cosmic rays are clearly subject to the laws of quantum physics and each cosmic ray particle has a range of possible paths to follow, only some of which give rise to the mutation. The mutation therefore fulfils the role of a measuring event, similar to the photon being detected by the polarizer. But if we consider the biological cell as a quantum system, we cannot say whether the mutation has occurred or not until we make a measurement on it. And if we go so far as to treat the whole planet as a quantum system, we cannot say that the species has evolved or not until we measure this. The world must retain the potential to behave both as if the species had evolved and as if it hadn't, in case a situation arises which brings these two possibilities together to reconstruct the original state in the same way as the 45° state is reconstructed by the reversed polarizer in Figure 4.2!

At this point the sensible reader may well be thinking something like 'Well, if quantum physics is saying that a gun can be half fired and half not fired at a cat who is then half dead and half alive, or that the world contains a biological species that half exists and half doesn't, then this is just ridiculous. I am going to put this book down and forget all this nonsense!' But it is the fact that these implications of quantum measurement theory are so absurd that is the main point of the argument. However successful quantum physics may have been in explaining the behaviour of atomic and subatomic systems, it should be clear by now that its statements about counters, cats and biological systems are quite wrong. What we hoped would be the final, fundamental theory of the physical universe is fatally flawed. But how is the theory to be modified to become acceptable? At some point in the measurement sequence the quantum description must become invalid, the chain must be broken and it must be possible to say that a physical system is in a particular state. We have seen that this point is beyond the stage of the single photon and polarizer, but whether the change occurs actually in the detector or at a later point is still an open question which has been the subject of considerable debate. It is the aim of the rest of this book to discuss the main themes of this debate. We shall be led to consider some amazing, strange ideas about the nature of the universe and our place in it. Are we unique creatures with souls and is this essential to any understanding of physical reality? Is there not one universe, but many that interact fleetingly during measurements? Or is there some more 'down-to-earth' solution to the problem? We start our discussion in the next chapter with a consideration of the first of these questions.

5 · Is it all in the mind?

We saw in the last chapter that the measurement problem in quantum theory arises when we try to treat the measurement apparatus as a quantum system: we need more apparatus to measure which state the first apparatus is in and we have a measurement chain that seems to go on indefinitely. There is, however, one place where this apparently infinite sequence certainly ends and that is when the information reaches us. We know from experience that when *we* look at the photon detector *we* see that either it has recorded the passage of a photon or it hasn't. When *we* open the box and look at the cat either it is dead or it is alive; *we* never see it in the state of suspended animation that quantum physics alleges it should be in until its state is measured. It might follow, therefore, that human beings should be looked on as the ultimate measuring apparatus. If so, what aspect of human beings is it that gives them this apparently unique quality? It is this question and its implications that form the subject of the present chapter.

Let us examine more closely what goes on when a human being observes the quantum state of a system. We imagine the usual set-up where a 45° photon passes through a polarization analyser which moves a pointer to one of two positions (H or V) depending on whether the photon is horizontally or vertically polarized. At least that is what would happen if the analyser and pointer behaved as a measuring apparatus; if on the other hand we treat them as part of the quantum system the pointer must be imagined as delocalized between H and V until its state is measured. We now add a human observer who looks at the pointer (it would be possible to imagine the observer using one of his other senses such as hearing a particular sound caused by a change in the pointer's position, but it is clearer if we think of a visual observation). In physical terms this means that light is scattered from the pointer into the observer's eyes, the signal is picked up by the retina and transmitted along the optic nerve to the brain. So far the process would seem to be just like that carried out by any other measuring apparatus and there would not seem to be any evidence of a uniquely human act. From now on, however, the measurement becomes part of the observer's knowledge. He is conscious of it. It is in his mind. The attribute of human beings that distinguishes us from other objects in the universe is our consciousness and, if we adopt this approach to the quantum measurement problem, consciousness has an even more central role to play in the physics of the universe than we might ever have imagined.

An example of the distinction between the conscious observer and a more conventional measuring apparatus is illustrated by a variation of the Schrödinger's cat situation known as 'Wigner's friend', after E. P. Wigner who was central in the development of the consciousness-based theory of measurement. In this example we replace the cat by a human 'friend', and the gun by a conventional detector and pointer. When we open the box, we ask our friend what happened and she will tell us that the pointer moved to H or V at a certain time. Assuming, of course, our friend to be a truthful person, we cannot now treat the whole box and its contents as a quantum system as the friend would then have to be in some state in which she didn't know whether the pointer was at H or V until we asked her! The cat may not really have been alive or dead, but the state of our friend's mind is quite certain – at least to her.

A consciousness-based quantum measurement theory therefore relies on the premise that human consciousness behaves quite differently from any other object in the universe. In the rest of this chapter we shall look at some of the evidence for and against this proposition and try to see whether or not it forms the basis of a satisfactory quantum measurement theory.

The idea that human consciousness is unique and different from anything else in the universe is of course a very old and widely held belief. Ever since men and women started thinking of their existence (which is probably ever since we were 'self-conscious') people have thought that their consciousness, sometimes called their 'mind', their 'self' or their 'soul', was something distinct from the physical world. This idea is a central tenet of all the world's major religions which maintain that this consciousness can exist independently of the body and indeed the brain – in some cases in a completely different (perhaps heavenly) existence after the body's death, and in others through a reincarnation into a completely new body or into the old body when resurrected at a Last Judgement.

A fairly recent exposition of the idea of the separateness of the soul is set out in a book written jointly by the famous philosopher Sir Karl Popper and the Nobel-prize winning brain scientist Sir John Eccles. The book's title *The Self and its Brain* clearly indicates the viewpoint taken. It is of course impossible to do justice to nearly 600 pages of argument in a few paragraphs, but it is possible to summarize the main ideas. Popper starts with a definition of 'reality': something is real if it can affect the behaviour of a large-scale physical object.* This is really quite a conservative definition of reality and would be generally accepted by most people, as will become clear by considering a few examples. Thus (large-scale) physical objects themselves must be real because they can interact and affect each other's behaviour. Invisible substances, such as air, are similarly real if only because they exert effects on other recognizably real solid objects. Similarly gravitational and magnetic fields must be real because their presence causes objects to move: dropped objects fall to the floor, the moon orbits the earth, a compass needle turns to point to the magnetic north pole and so on. All such objects, substances and fields are described by Popper as belonging to what he calls 'world 1'. There are two more 'worlds' in Popper's philosophy. World 2 consists of states of the human brain which may well be describable in terms of the patterns of electrical impulses on the neurones. These brain states must be considered real for exactly the same reasons as were world 1 objects, that is they can affect the behaviour of physical objects. Thus a particular state of the brain can cause a message to be transmitted along a nerve that causes the contraction of a muscle and a movement of a hand or leg which in turn may cause an undoubted world 1 object such as a football to be propelled through the air.

Beyond worlds 1 and 2 is world 3. Following Popper, world 3 is defined as the products of the human mind. These are not physical objects, nor are they merely brain states, but are things such as stories, myths, pieces of music, mathematical theorems, scientific theories etc. These must all be considered real for exactly the same reasons as were applied to worlds 1 and 2. Consider for example a piece of music. What is it? It is certainly not the paper and ink used to write out a copy of the score, neither is it the gramophone record on which a particular performance is recorded. It isn't even the set of sound vibrations in the air when the music is played. None of these world 1 objects *are* the piece of music, but all exist in the form they do *because of* the music. The music is a world 3 object, a product of the human mind, which is to be considered as 'real' because its existence affects the behaviour of large scale physical objects – the ink and paper of the score, the shape of the grooves in the record, the pattern of vibrations in the air and so on.

* Popper actually refers to large-scale objects partly to avoid discussing the quantum behaviour of microscopic bodies, so there is a potential problem of consistency here if we are to apply his ideas to the measurement problem where, as we have seen, the quantum behaviour of large-scale bodies is important.

Another example of a world 3 object is a mathematical theorem such as 'The only even prime number is 2'. Everyone who knows any mathematics must agree that this statement is true and it follows that it is 'real' if only because world 1 objects such as the paper of this page and the arrangement of the ink on it would have been different otherwise. The reality of scientific theories is real in even more dramatic ways. It is because of the truth of our scientific understanding of the operation of semiconductors that micro-chip-based computers exist in the form they do. Tragically, it was the truth of the scientific theories of nuclear physics that resulted in the development, construction and detonation of a nuclear bomb.

The reader may well have noticed an important aspect of these world 3 objects. Their reality is established only by the intervention of conscious, human beings. The piece of music or the mathematical theorem results in a particular mental state of a human being (i.e. a world 2 object) which in turn affects the behaviour of world 1. Without human consciousness this interaction would be impossible and the reality of world 3 could not be established. It is this fact that leads Popper and Eccles to extend their argument to the reality of the self-conscious mind itself. Only a self-conscious human being can appreciate the reality of world 3 objects, which are real because their existence can (via human consciousness and brain) affect world 1 objects. It follows that human consciousness itself must be real and different from any physical object, even the brain.

These ideas are developed further in a major section of the book, written by John Eccles. He describes the physiological operation of the brain and speculates on how the self-conscious mind may interact with the brain: he suggests a remarkably mechanistic model in which he postulates the existence of particular 'open synapses' in the brain that are directly affected by the (assumed separate) self-conscious mind. An interaction of this kind is a necessary consequence of the idea of a mind or soul that is separate from the body and brain: before the, undoubtedly real, world 1 events can occur there must be an interaction between the 'thoughts' of the mind and the physical states of the brain. At some point there must be changes in the brain that do not result from normal physical causes but which are the result of a literally 'supernatural' interaction.

The above arguments are by no means universally accepted and many people (including the present author) believe that a much more 'natural' understanding of consciousness is possible. But if we accept the idea of our conscious selves as separate from and interacting with our physical brains, a resolution of the measurement problem is immediately suggested. We simply postulate that the laws of quantum physics govern the whole of the physical universe and that the measurement chain is broken when the information reaches a human consciousness. The interaction between mind and matter, which by definition is not subject to the laws of physics, breaks the measurement chain and puts the quantum system into one of its possible states.

The effect of this view of the quantum theory of measurement on our attitude to the physical universe can hardly be exaggerated. Indeed it is difficult to hold this position while still assigning any reality at all to anything outside our consciousness. Every observation we make is equivalent to a quantum measurement of some property which apparently has reality only when its observation is recorded in our minds: if the state of a physical system is uncertain until we have observed it, does it mean anything to say that it even exists outside ourselves? 'Objective reality' (the reality of objects outside ourselves) seems, in Heisenberg's phrase, to have 'evaporated' as a result of quantum physics. As Bertrand Russell put it in 1956: 'It has begun to seem that matter, like the Cheshire Cat, is becoming gradually diaphanous and nothing is left but the grin, caused, presumably, by amusement at those who still think it is there'. Of course the existence of the external universe has always been recognized as a problem in philosophy. Because our knowledge of the outside world (if it does exist!) comes only through our sense impressions, it is only this sensual data of whose existence we can be sure. When we say, for example, that there is a table near us, all we actually know is that our mind has acquired information by way of our brain and our senses that is consistent with the postulate of a table. Nevertheless, before quantum physics it was always possible to argue that by far the simplest model to explain our sense data is that there really is a table in the room – that the external physical universe does exist. A quantum theory based on consciousness, however, goes further than this: the very existence of an external universe, or at least the particular state it is in, is strongly determined by the fact that conscious minds are observing it.

We have reached a very interesting position. Ever since the beginnings of modern science four or five hundred years ago, scientific thought seems to have moved man and consciousness further from the centre of things. More and more of the universe has become explicable in mechanical, objective terms, and even human beings are becoming understood scientifically by biologists and behavioural scientists. Now we find that physics, previously considered the most objective of all sciences, is reinventing the need for the human soul and putting it right at the centre of our understanding of the universe! However, before accepting such a revolution in attitudes, it is important to examine some of the arguments against a consciousness-based measurement theory and to explain why, although some continue to support it, most physicists do not believe it is an adequate solution to the measurement problem nor, indeed, a correct way to understand the physical universe and our relationship with it.

The main problem with countering a subjective philosophy is in its obvious self-consistency. The basic assumption that the only information we can have about everything outside ourselves is a result of sense impressions (apart, presumably, from *extra*-sensory perception or divine revelation, if such things exist!) is incontestable. It follows that it is impossible ever to prove the existence of the external world. However, there are a number of important arguments which make a purely subjective view in which the physical world has no objective existence and our consciousness is the only reality appear unreasonable, at least. Perhaps the most important of these is that different conscious observers agree in their description of external reality. Suppose a number of people driving cars approach a set of traffic lights: if they did not agree with each other about which light was on and what its colour was, a catastrophic accident would certainly result. The fact is that all these drivers experience the same set of sense impressions which they all attribute to the objective existence of a red light, so that they all stop, or of a green light, so that they all keep going. Now it is perfectly possible to argue that, by some coincidence, their brains and consciousness are all happening to change in similar ways and that the light has no real objective existence, but such an explanation is complex to the point of being perverse compared with the simple objective statement that a light really exists. The extreme, if not the logical, conclusion of subjectivism is to believe that the information received from other conscious beings also has no reality, but is part of one's own sense impressions. Thus my own (or your own?) thoughts are the only reality and everything else is illusion. There is only me: not only the car and traffic lights, but also the other drivers and their consciousnesses are figments of my imagination. Such a viewpoint is known as solipsism and, by its very nature, puts an end to further discussion about the nature of reality or anything else. If everything, including this book and you reading it, are just figments of my imagination there wouldn't seem to be much point in my writing it; on the other hand if this book and I are just figments of your imagination there wouldn't seem to be much point in your reading it!

We have therefore been led to reject a purely subjectivist philosophy, not because it can be proved inconsistent, but because its consequences lead us to statements that, although not disprovable, are complex and completely unreasonable. A test of simplicity and reasonableness has always formed an important part of scientific theory. It is always possible to invent over-elaborate models to explain a set of observed facts, but the scientist, if not the philosopher, will always accept the simplest theory that is consistent with all the data. There is a (no doubt apocryphal) story about a person who always spread salt on the floor before going to bed at night. The reason for doing so was 'to keep away the tigers'. When told that no one had ever seen a tiger in this part of the world the reply was 'that shows how cleverly they keep out of sight and what a good job the salt is doing'. An important test of any scientific theory is that it should have no 'tigers', i.e. no unnecessary postulates. The difficult with theories of quantum measurement is that they all appear to contain 'tigers' of one kind or another and there is no general agreement about which theory contains the greatest number or the fiercest ones! The point of the last few paragraphs has been to show that a theory based on the idea that our subjective consciousness is the only reality is a tiger of the fiercest, or even man-eating kind!

But just a minute. It's all very well, and all too easy in fact, to criticize the idea that consciousness is the only reality, but is this really what a consciousness-based theory of quantum measurement is saying? The mind may play a crucial role in the measurement process, even to the point that a choice between possible quantum states is made only when the signal is recorded on a consciousness, but the *existence* of the physical system need not therefore be in doubt. Even if all the properties of the physical system are quantum in nature, in the sense that their values are attained only when they are observed by a conscious observer, the possible outcomes of these observations are quite outside the control of the observer. The traffic lights can only be red, green or amber – no one can turn them blue or purple just by looking at them. The photon emerging from the HV polarizer is seen to be either horizontally or vertically polarized by any conscious observer: no one can change the polarization to 45° or double the number of photons passing through the apparatus just by observing it. Is it not possible to maintain the idea of consciousness as the end of the measurement chain without going anywhere like as far as saying that subjective experience is the only reality?

The difficulty with answering yes to the above question is to draw a distinction between the existence of an object – be it photon, measuring apparatus or 'world 3' concept – and its properties. If all the properties of an object, its mass, position, energy etc., are quantum in nature and do not have values until they are measured, it is hard to see any meaning in the object's separate existence. Over and above this, however, the consciousness-based theory of measurement still leads to some conclusions which are incredible, to say the least, and which correspond to 'tigers', just about as large and ferocious as some of those encountered earlier. A consciousness-based quantum measurement theory states, in brief, that the choice of possible states of a quantum system and its associated measuring apparatus is not made until the information has reached the mind of a conscious observer: the cat is neither alive nor dead until one of us has looked into the box; the species both evolved and didn't evolve until observed by a conscious person. Is it reasonable to think that the presence or absence of a biological species today, and of its fossil record over thousands of years were determined the first time a conscious human being appeared on the planet to observe them? Such a view is surely hardly more credible than the suggestion that all reality is subjective.

Such objections have led some thinkers to suggest that consciousness is not just a property of human beings, but is possessed to a greater or lesser extent by other animals (in particular cats!) and even inanimate objects. Alternatively, others have suggested that the world is observed, not only by ourselves, but by another eternal conscious being, whom we might as well call 'God'. The idea that God has a role in ensuring the continual existence of objects that are not being observed by human beings is actually quite an old one and led to the following nineteenth-century limerick:

There once was a man who said, 'God
Must think it exceedingly odd
If he finds that this tree
Continues to be
When there's no one about in the quad'

and its reply

Dear sir, your astonishment's odd
I am always about in the quad
And that's why the tree
Will continue to be
Since observed by, yours faithfully, God

A similar idea has been stated more prosaically by a modern writer on quantum measurement problems who writes

If I get the impression that nature itself makes the decisive choice what possibility to realize, where quantum theory says that more than one outcome is

possible, then I am ascribing personality to nature, that is to something that is always everywhere. Omnipresent eternal personality which is omnipotent in taking the decisions that are left undetermined by physical law is exactly what in the language of religion is called God.
F. J. Belinfante; *Measurements and Time Reversal in Objective Quantum Theory*, Pergamon, 1975.

The difficulty with this point of view is that it does little to solve the problem, but merely restates it. If everything has consciousness or if God's consciousness is determining which state a quantum system will occupy, then we are still left with the question of at which point in the measuring chain this choice is exercised. Presumably God doesn't look at the photon passing through the polarizer at least until the detector state has changed. Why not? And if not, why not until the information has reached a *human* consciousness. We are simply back where we started, not knowing at what point the measuring chain ends and why. Turning Belinfante's ideas around, the idea of God choosing is no different, and certainly no more satisfactory than that of nature choosing.

This is not of course to say that God cannot exist, but only that this idea does not help us to solve the quantum measurement problem. Similarly, although we have seen that a consciousness-based measurement theory leads to unacceptable consequences, we have not thereby disproved the existence of consciousness or the soul. Indeed many people, including some scientists and philosophers, continue to believe in God and in the human soul without caring one way or the other about quantum theory, and most of Popper's arguments for the existence of 'world 3' objects and of the mind are quite untouched by what we have said so far. Nevertheless, and although it is not strictly relevant to the quantum measurement problem, we shall give a brief outline of some of the modern ideas about consciousness and the brain that argue against the idea of a separate mind or soul.

Let us first say that Popper's arguments, summarized earlier in this chapter, for the reality of 'world 3' objects and consciousness are probably correct and certainly convincing. If a real object is something that can cause a change in a large-scale material object then 'world 3' objects and consciousness are indeed real and do transmit their effects through 'world 2' states of the brain. Where many people part company with Popper, and particularly with his co-author John Eccles, is when it is suggested that consciousness is separate from the brain – that there is 'a ghost in the machine'. A modern view of the relationship between consciousness and the brain is to draw an analogy with the relationship between a computer program and a computer. A computer is a complex array of electronic switches with no pattern or purpose in itself. It is only when it is programmed – i.e. when the switches are made to operate in a particular sequence – that the computer operates in a useful manner. On the other hand, the program has no existence apart from the computer and is certainly not independent of it in a 'soulist' sense. In the same way it is possible that the mind, although real in the same way as the computer program is real, is not separate from the brain in the same way that the program is not separate from the computer. To use modern jargon, the program (consciousness) is 'software' while the computer (brain) is 'hardware'. This view of consciousness receives some support from research into 'artificial intelligence' where the capacity of programmed computers to 'think' is investigated. Already some computers can play chess nearly to Master standard, can answer questions in an apparently intelligent way and, when fed with the appropriate information, can recognize human faces. All experts agree that this is a long way from behaving anything like a fully conscious human being, but at the present rate of progress it may not be many years before a computer is built and programmed so that its behaviour is indistinguishable from that of a conscious human mind. Of course this may still prove impossible and, even if it is achieved, some will still argue that the programmed computer is not self-conscious in the same way as a human being is. However, the possibility that we will eventually be able to understand consciousness in this way is now so real that the basing of a quantum theory or philosophy on the existence of consciousness as a unique separate non-physical entity must be considered doubtful to say the least.

Quantum physics and ESP

Before closing this chapter on the consciousness theory of measurement, we discuss briefly the suggestion, often made in some circles, that the ideas of quantum physics can be used to at least partly explain so called 'paranormal' phenomena associated with 'extra-sensory perception' and so forth. It is important to emphasize that we are going to discuss, not the evidence for or against the existence of these effects, but only whether quantum physics can reasonably be thought to give some scientific support and respectability to such ideas.

A typical ESP situation might be where one experimenter sends messages about something like the pattern on a card to another person in a completely different room, with no known communication between them. After this has been repeated a large number of times with many cards, a success rate considerably greater than that allowed by chance is sometimes claimed. Looking at this from the quantum point of view, we might first be struck by a superficial connection with the EPR experiments discussed in Chapter 3. We saw there that the correlations between the polarizations of widely separate photons were greater than could be accounted for from correlations built in at the time the pair was created and we might be led to postulate that there are correlations between the minds of the separate experimenters for similar reasons. But this would be to ignore the crucial arguments set out at the end of Chapter 3 which show that the EPR apparatus cannot be used to pass signals from one polarization apparatus to the other. Moreover, although we cannot explain the correlations simply by assuming that they were built in when the pair was created, the photons are still needed before the correlations occur. The quantum predictions (confirmed by the Aspect experiment) are predictions about pairs of photons created with these particular properties. There is nothing analogous to the photon pair in the ESP case and no way in which quantum physics would allow the application of EPR-type reasoning in this situation.

Another way in which it is sometimes suggested that quantum physics can be related to ESP is through a consciousness-based measurement theory. If the 45° photon retains its potential for both H and V polarization until it is observed by a self-conscious mind, if the cat is both alive and dead until someone looks at it, then mind is apparently influencing matter. Are we then in a position to explain psychokinesis in which some conscious minds with particular powers are said to be able to cause objects to move around rooms or to bend spoons or whatever? At a less dramatic level, is it unreasonable to suggest, as has been claimed, that the conscious observer could influence the time at which a radioactive atom decays? Despite the superficial attractiveness of such ideas it should be clear that these alleged phenomena are no more consistent with quantum theory than they are with classical physics. This follows because, even if the mind is the final (or the only) measuring apparatus, *it acts as a measuring apparatus*. It is true that in quantum physics the observed system is influenced by the measurement, but this influence is limited to determining the nature of the possible outcomes of the experiment, –what we called first-level operations in Chapter 4 – and therefore cannot affect the 'second-level' statistical results which are all precisely and correctly predicted by quantum theory. Whether or not it is the mind that is finally responsible for the measurement, if a large number of 45° photons have passed through an HV apparatus, *either* 50% of them have emerged in each channel *or* the laws of quantum physics have been violated.

It should be emphasized that we have attempted to argue, not the truth or falsity of the existence of extra-sensory perception and related phenomena, but only that we cannot appeal to quantum theory to make them more reasonable or acceptable. Even a consciousness-based quantum measurement theory ascribes quite a different role to the mind than that required in this context and if such phenomena were to be established with the same reliability and reproducibility as is exhibited by, say, the photon pairs in an Aspect experiment, they would require an explanation that is right outside present scientific ideas, either classical or quantum.

6 · Many worlds

A completely different interpretation of the measurement problem, one which many professional scientists have found attractive if only because of its mathematical elegance, was first suggested by Hugh Everett III in 1957 and is known variously as the 'many-worlds' or the 'branching-universe' interpretation. This viewpoint gives no special role to the conscious mind and to this extent the theory is completely objective, but we shall see that many of its other consequences are just as revolutionary and bizarre in their own way as those discussed in the previous chapter.

The essence of the many-worlds interpretation can be illustrated by again considering the example of the 45° polarized photon approaching the HV detector. Remember what we demonstrated in Chapter 4: from the wave point of view a 45° polarized light wave is equivalent to a horizontally polarized wave and a vertically polarized wave added together and, if we were able to think purely in terms of waves, the effect of the HV polarizer on the horizontally polarized wave would be simply to split the wave into these two components with one going along each HV channel. The many-worlds interpretation of quantum mechanics applies this simple idea to the photon: instead of the photon following only one path or the other it splits and follows both. However because we can't have half a photon, this split is actually a transformation from a single 45° photon into a *pair* of photons polarized in the horizontal and vertical directions. But just a minute. If a photon actually splits amoeba-like into two photons, how is it that only one of the pair is ever observed? If we can multiply photons in this way, we could surely create an indefinite amount of light out of nothing!

To look after this point, we remember what happens when we put a photon detector in the beam to record in which HV channel the photon emerges. We saw in Chapter 4 that the detector has two possible pointer states, corresponding to the HV states of the photon. In the same way as the HV measurement causes the photon to split into two channels, so an observation of the measuring apparatus will cause this to split into two also! And so on along the measuring chain. Even if there is a human observer in the chain he or she is split into two, each one of whom is aware of the photon in one or the other channel. 'Now this is really ridiculous', I hear you say, 'If measuring apparatus, not to mention people, were splitting and multiplying in this way I would certainly have heard about it, or noticed it when it happened to me!' This leads us to an absolutely crucial point about the many-worlds interpretation: *once a split has occurred, the two branches have no way of being aware of each other*. Thus when the 45° photon is split in the HV apparatus there is no interaction between the two photons so created, and the same is true when a measuring apparatus or even a person divides. The two measuring apparatuses or people continue to exist, but they are completely unaware of each other's existence. It follows that because each measuring apparatus is interacting with the world about it and indeed eventually with the rest of the universe, each component of the split observer must in fact carry along with him a copy of the entire universe. Hence the term 'branching universe': whenever a quantum measurement occurs the universe branches into as many components as there are possible results of the measurement. Everyone in a particular branch thinks that their result of the measurement and their particular universe is the only one that exists. The different branches of the universe almost never come together again.

We say 'almost never' in the last paragraph because a reunion of different branches is sometimes possible. Consider the case of the 45° polarized photon that passes through the HV apparatus and then through a crystal that reunites the two possible channels to recreate the original 45° polarized state in the manner discussed in Chapter 4. From the many-worlds point of view the photon splits into two while passing through the apparatus, and the two photons reunite in the crystal. Suppose now that there are detectors in the H and V beams. As the two photons enter their detectors, the detectors and everything they are interacting with are split and we would soon have created two new branches of the universe. But suppose, before the interaction has gone any further, we imagine that the two split photons are brought together and simultaneously the original detector states are reconstructed. We saw in the last chapter that it is just this type of operation (impossible in practice, but not in principle) that leads to the whole measurement problem. From the many-worlds point of view this would correspond to a merger of the two split detectors and photons into single entities just as they were initially. If a human observer were to be present and split and if we imagine that his state is also reconstructed so as to be just the same as it was before the measurement, then clearly all memory of the split states would have been destroyed. We could even imagine that some time after the whole universe has been split into two by a measurement, the two halves could subsequently evolve to the point where they matched each other accurately enough to allow recombination to take place. But even if such an incredibly improbable event did occur, it could do so only by washing out all memory of its condition during the time it was split.

The postulates of the many-world interpretation may be incredibly radical, but this is at least partly compensated for by the resolution of many of the puzzles and paradoxes of the measurement problem. Consider the theory applied to Schrödinger's cat. Instead of having to worry about how a cat can be neither dead or alive, we simply have two cats, one alive and one dead, each in its own box in its own universe, and two observers each opening their own box and drawing their own conclusions about the state of the cat and hence of the photon. Or think of Wigner's friend. We no longer have to worry about whether the measurement is made when the friend observes the photon or when Wigner learns the result: as the photon passes through the measuring apparatus it is split and so are Wigner, friend and all. In one universe the friend sees a horizontally polarized photon and tells Wigner this result. In another she sees a vertically polarized photon and passes this message on.

One of the most important features of the many-worlds picture is the fact that its interpretation seems to arise naturally out of the mathematical formalism, whereas the more conventional approaches require additional assumptions associated with the distinction between the quantum system and the measurement apparatus. It is difficult to appreciate this point without going into more mathematical detail than is appropriate for a discussion at this level, but some understanding may be gleaned by considering the situation depicted in Figure 6.1. We imagine four separate sets of apparatus (including detectors) for measuring HV polarization; we allow a photon to pass through one of the sets of apparatus and we assume, as usual, that each photon is in a state of 45° polarization before the measurement. We then repeat the experiment with other photons and the other three sets of apparatus and we consider what the branching universe idea has to say about this sequence of operations. Initially there is only one universe in which none of the four detectors has registered a photon: this is described by the symbol OOOO at the left of the network in Figure 6.1. Suppose now that one of the photons passes through one of the polarizers; this has the effect of splitting the universe into two, one of which has its detectors in

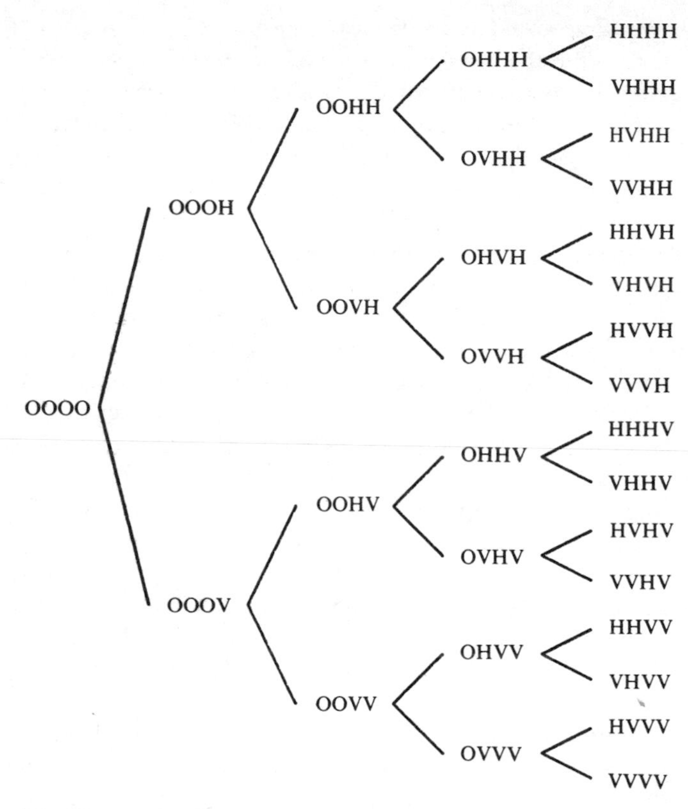

Fig. 6.1 We consider, from the many-worlds point of view, four separate measurements of HV polarization of 45° photons. Beforehand all four measuring apparatuses are in their initial state O. After the first measurement the universe splits into two branches: in one branch a photon and measuring apparatus are in the state H while in the other they are in the state V. Further splittings follow at each measurement and, when the process is completed, the results are approximately equally divided between H and V in most universes.

the state OOOH and the other in the state OOOV. We continue passing the photons one at a time until all four have been measured. We now have a total of sixteen universes representing all the possible permutations of H and V as shown on the right of Figure 6.1. If we now examine how many times H and V photons were recorded in a particular universe we find that six universes recorded 2H and 2V, four recorded 1H + 3V, another four 1H + 3V and in only one universe each were 4H or 4V observed. The important point is that this is just the same conclusion as we would have reached from the standard analysis of this experiment as described in Chapter 2: a 45° photon passing through an HV polarizer is expected to have a 50% chance of emerging through either channel, but if there are only four photons, deviation from this ideal behaviour is quite likely. If the number of photons is increased, the distribution should get nearer and nearer the ideal 50-50 division, and this is also found to happen in the branching-universe model. For example a network similar to that shown in Figure 6.1, but with ten photons and ten polarizers, produces 1024 universes, 622 of which are within 10% of the ideal 50-50 distribution while the states 10H and 10V have only one universe each.

It must be admitted that choosing a case where the outcome is 50-50 has considerably simplified the above argument. To handle a more general case like the measurement of HV polarization on a photon initially polarized in a general direction is more difficult and requires a further assumption. However this arises quite naturally from the mathematical framework of the theory and, if a large number of experiments are considered, the branching-universe theory leads to the same probabilities for the various possible outcomes as are obtained by conventional methods.

The simplification of the measurement problem achieved by the many-worlds approach, combined with the fact that the mathematical details of the theory turn out to be very elegant, has led the physicist Paul Davies to describe it as 'cheap on assumptions, but expensive on universes!' Before trying to decide whether the net balance of this account is profit or loss, let us look at a few more of the consequences of the many-world assumption.

We might wonder how many universes there are. As every quantum event has two or more possible outcomes and as huge numbers of branching events are continually occurring, the number must be quite immense. It has been estimated that there are about 10^{80} elementary particles in the observable part of any one universe. If we assumed that each of these had been involved in a two-way branching event each second in the 10^{10} years since the 'big bang', the number of universes created by the present time would be something like $10^{10^{12}}$, an unimaginably huge number, and this is certainly a lower estimate.

Where are all these universes? The answer is that they may all be 'here' where 'our' universe is: by definition universes on different branches are unable to interact with each other in any way (unless they are able to merge in the very special circumstances mentioned earlier) so there is no reason why they should not occupy the same space.

Alternatively we can imagine the universes stacked up in some extra dimension of space we know nothing about. As the other universes, some of which include other copies of ourselves, can almost never interact with the branch we are on, and even if they do they cannot carry forward any information about their split state, we can never find the answer to such questions.

The possibility of the existence of other universes some of which differ only slightly from our own leads to some enticing speculations. Thus there would be many universes which contain a planet just like our own earth, but in which a particular living species is absent, because the quantum event that caused the mutation which led to the development of the species in our universe (and many like it) did not follow this path in these other universes. Presumably there are universes in which life evolved on earth, but man didn't, and which are therefore pollution-free and not threatened by imminent nuclear extinction. If we consider that every choice between possible outcomes is fundamentally a result of a quantum event, then every such possibility must exist in its own set of universes. Every choice we have made in our lives may be associated with a quantum event in our brains, and if so there would be universes with other versions of ourselves acting out the consequence of all these alternative thoughts! Some of the other universes will be delightful places: because of a happy coincidence of quantum events our favourite football team wins every trophy every year and society is constructed the way we would like it because our favourite political party wins every election. In others, of course, the opposite happens and in many the possibility of irrational political behaviour resulting in the extinction of mankind in a nuclear confronatation must already have been realized.

We might imagine we could test the many-worlds hypothesis by performing an experiment in which one of us was split by a quantum measurement and reunited again. The person who underwent this experience could then tell us what it felt like to split and could remember having observed the photon in each of its polarization states. The problem with this idea is that, as pointed out earlier, the recombination of the two halves is possible only if at the same time all information about which channel was followed is erased in the same way as the undetected photon has its original 45° polarization restored by the reversed calcite crystal. It follows that if a human observer is involved any memory of the outcome of the experiment must also be erased. Indeed it used to be generally believed that reunification would be impossible unless the observer were restored to the same state as before the experiment, so that any memory of the split would also be erased. Recent theoretical work by Professor David Deutsch indicates that this may not be correct and that if such an experiment could be carried out an observer could tell us what the split felt like, but not of course what the result of the experiment was. He has suggested that if a sufficiently intelligent computer were built sometime in the future, it could play the role of the observer in such a process. If, indeed, human consciousness is different from a computer in not being subject to many-worlds splitting, this could conceivably be tested experimentally by comparing human and computer experience in such a situation.

Despite all we have said, it would be surprising if many readers have been convinced of the plausibility of many-worlds theory. In the vast majority of the universes that have evolved since you began reading this chapter, you are completely sceptical about the whole idea! To postulate the existence of a near-infinite number of complete universes to resolve a subtle theoretical point seems to be introducing 'tigers' into our thinking with a vengeance. We have seen that the other universes do not interact with us except possibly in very special circumstances when no information is obtained about them in any case. In which case, does the statement that they exist mean anything? In his original paper on the many-worlds interpretation Everett compares such criticisms with the objections of the mediaeval world to Copernicus's suggestion that the earth revolves around the sun. In the same way that people were prejudiced by their entrenched belief that the earth is the centre of the universe, it is suggested that we find the branching-universe idea unacceptable simply because we have been brought up with the prejudiced idea that only one universe exists. He would say that in each case acceptance of the radical postulate allows us to make sense of experimental results that were previously illogical and confused. Nevertheless most scientists would not agree that the two cases are parallel and would maintain that the extravagance of the original postulate is much greater in the many-worlds case and the gain in clarity of understanding is much less. To adapt Paul Davies's words, most would believe that the losses involved in the 'extravagance with universes' heavily outweigh the gains from the 'economy with postulates'.

There are other rather more technical reasons for doubting the power of the many-worlds model. These arise principally because it is not clear from the theory just when the alleged branching takes place. It is sometimes said that it happens whenever a 'measurement-like' interaction between a quantum system and a measuring apparatus occurs, but if this is the case the many-worlds view has signally failed to solve the measurement problem! The main difficulty with the Copenhagen interpretation is the need to make a distinction between the system and the apparatus and if this is not resolved, or at least avoided, by the many-worlds theory the 'economy with postulates' is beginning to look rather small. Alternatively, branching may occur whenever any kind of interaction takes place between two component parts of the universe. Does this mean that the electron and proton in a hydrogen atom are continually interacting and creating infinities of universes or is it only when particles interact briefly as in a scattering experiment that splitting occurs? The many-worlds formalism does not seem to give clear answers to such questions.

It is instructive to consider the application of the many-worlds model to the EPR situation in which the polarization of the separated components of correlated photon pairs are measured (see Chapter 3): whenever one photon emerges in (say) the horizontal channel of an HV apparatus its partner behaves from then on as if it were vertically polarized. Adopting the first version of the many-worlds theory we would say that the measurement of the HV polarization on, say, the left-hand photon causes a splitting into two universes: one of these has a horizontal photon on the left and a vertical photon on the right and in the other both polarizations are reversed. But this has certainly not overcome the problem of non-locality. The Aspect experiment has shown that the correlations between photon properties are propagated instantaneously (or certainly faster than light) so any splitting of the universe must also happen at the same rate if it is to account for the experimental results. If a measurement performed on a photon at one place causes an immediate splitting of the whole universe out to the furthest stars, a non-locality is implied that is certainly as radical as, and probably more so than, any other version of quantum theory. Problems also arise if we adopt the second version of many-worlds theory and postulate a splitting of the universe at the time the photon pair is created. We now have to postulate that a separate universe exists corresponding to every possible orientation of the photon polarization. Consider one of these universes containing an HV apparatus; it is extremely unlikely that this universe will also contain a photon pair that is horizontally or vertically polarized, so it follows that when one of the photons arrives at the HV apparatus a further splitting must occur which must propagate throughout the universe instantaneously for the same reasons as discussed earlier.

We can conclude that the many-worlds postulate certainly does not resolve the problem of non-locality and it is difficult to see how it overcomes the measurement problem. What it does do is recover some kind of realistic description of quantum objects. We can preserve the simple model of a photon following a particular path and having a particular polarization at the same time as keeping a record (in another universe) of what would happen if another path were followed. Even a form of determinism is recovered because all possible outcomes of a quantum event not only can, but do, occur. We mentioned in Chapter 1 how Laplace in the nineteenth century summed up the deterministic idea in the statement that 'the present state of the universe is the result of its past and the cause of its future'. In a similar way, the many-worlds physicist believes that the state of the particular universe we happen to be in is 'the result of its past' and is the 'cause of the future' of all the universes that will branch out from this one. In principle (but not of course in practice) the future states of all these universes could be calculated from the quantum laws and the present state of our own branch. However, it is interesting to note that it is impossible, even in principle, to make the same calculation about the past as this would require knowledge of the state of presently existing branches other than the one we happen to occupy.

Of course if we reject the branching-universe model, we are back again with the measurement problem. If the preservation of all possible states in a multiplicity of universes is unacceptable, then a choice out of the possible quantum paths must be made somewhere. And if it is unacceptable that this choice be made only in the conscious mind it must be made somewhere else in the measuring chain. Is there a point in the process, outside the human mind, at which we can say that the chain is broken and the measurement is complete? The next chapter explores this question further and later chapters suggest a possible basis for an answer.

7 · Is it a matter of size?

The last two chapters have described two extreme views of the quantum measurement problem. On the one hand it was suggested that the laws of quantum physics are valid for all physical systems, but break down in the assumed non-physical conscious mind. On the other hand the many-worlds approach assumes that the laws of physics apply universally, and that a branching of the universe occurs at every measurement-like situation or at some other point in the quantum process. However, although in one sense these represent opposite extremes, what both approaches have in common is a desire to preserve quantum theory as the one fundamental universal theory of the physical universe, able to explain equally well the properties of atoms and subatomic particles on the one hand, and detectors, counters and cats on the other. In this chapter and the two following we explore the alternative possibility that this may not be the case: that despite its apparent universality, quantum theory may not be applicable to large-scale macroscopic objects in the same way as it is to microscopic phenomena and that a resolution of the measurement problem may lie in a completely new way of looking at the relationship between the physics of large and small systems.

The first point to be made is that the problems we have been discussing seem to make very little difference in practice. As we emphasized in Chapter 1, quantum physics has been probably the most successful theory of modern science. Wherever it can be tested, be it in the exotic behaviour of fundamental particles or the operation of the silicon chip, quantum predictions have always been in complete agreement with experimental results. How therefore have working physicists resolved the measuring problem? The answer is quite simply that in practice physicists apply the Copenhagen interpretation, knowing perfectly well when a measurement has been made and what the distinction is between a quantum system and a measuring apparatus. A photon detector such as a photomultiplier tube with its high-voltage supply and its electronic circuits could never in practice be analysed in detail by quantum calculations, and physicists have no practical difficulty distinguishing between 'measurement' processes such as a particle being detected by such an apparatus and 'pure quantum' processes like a photon passing through a polarizer without being detected. Nevertheless if asked to define this distinction in a consistent way most of us would find it hard to define clearly where the dividing line comes. If we are to resolve the measurement problem at this point, we will have to find a means of making this distinction in a consistent manner. It can't be just a matter of the physical dimensions of the measuring apparatus because we saw in our discussion of the EPR problem in Chapter 3 that quantum correlations can be observed over large distances (several metres in the Aspect experiment). What other property clearly marks the distinction between a measuring apparatus and a quantum system on which measurements are made? In the present chapter we explore the possibility that the key factor may be the fact that the measuring apparatus consists of a large number of atomic particles, while in the next two chapters we consider whether the distinction may lie in the nature of the measuring process.

The most obvious feature distinguishing measuring instruments such as photomultiplier tubes, counters and human beings from pure quantum systems is that the former contain huge numbers of atoms that themselves consist of electrons, protons and neutrons while the latter typically consists of a single particle such as a photon or, at most, a very small number of particles. Moreover we can explain the behaviour of measuring instruments, certainly at the point where the information is processed to the stage where we can read it, in largely classical terms. Thus pointers in meters move across dials because electric currents are passing through coils which then are affected by magnetic fields; or lights are switched on and the emitted light is intense enough to be treated as a classical wave ignoring its photon properties. Could it simply be that, once objects contain a sufficiently large number of atoms they obey the laws of classical physics rather than quantum mechanics? Microscopic bodies would then obey quantum laws until they interacted with large bodies when a measurement interaction would take place.

One difficulty in this approach is to define just what is meant by 'large enough' or 'sufficiently many atoms' in this context. However, a typical measuring instrument might contain 10^{23} atoms which is an awful lot more than one or two and it is quite conceivable that the truly fundamental laws of physics contain terms we are presently unaware of that have a negligible effect when there are only a few particles in a system, but become appreciable when the number of particles is large. Thinking along these lines has led to a reappraisal of the evidence for the quantum behaviour of macroscopic systems.

It has long been realized that, although macroscopic bodies may appear to behave classically, many of their properties result from the quantum behaviour of the system's microscopic components. For example, the fact that some substances are metals which conduct electricity while others are insulators or semiconductors results from the quantum properties of the electrons in the solid. Similarly the thermal properties of a solid, such as its specific heat or its thermal conductivity, depend on the quantum motion of the atoms in it. It might be thought therefore that the quantum behaviour of macroscopic systems is a well established fact. However, there is one feature of the measurement process that is not reflected in such considerations. This is that when a record is made of a measurement the states of a large number of particles is changed together. For example, when a pointer moves across a dial, all the atoms move together; or when a light is switched on it is the collective movement of the electrons in the wire which gives rise to the electric current that lights the bulb. Such cooperative motion of a large number of particles is an essential feature of all observable macroscopic events and of measuring instruments in particular. Could it be that quantum theory breaks down just when such large-scale collective motions are involved? To test the applicability of quantum physics in such a situation we might consider using an apparatus like that illustrated in Figure 7.1: here we have the usual example of a 45° photon entering an HV apparatus the output of which is connected to a detector that swings from the central position O to the points H or V depending on which channel the photon emerges through. To this extent the apparatus is

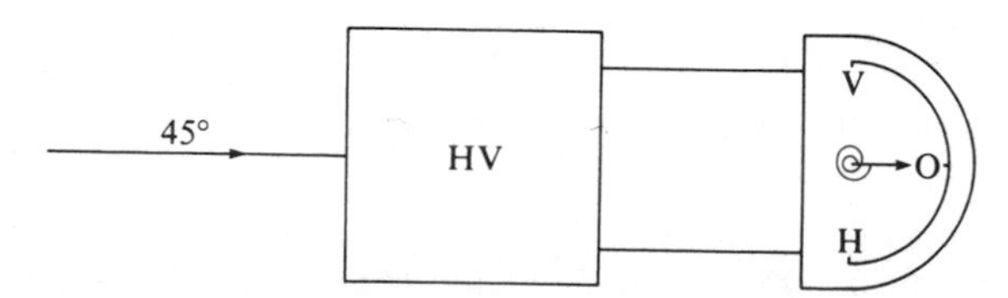

Fig. 7.1 If the pointer on a photon detector is mounted on a perfect spring, will it stand still because the motions induced by the horizontal and vertical signals cancel each other? Or does the pointer always move to either H or V because quantum physics cannot be applied to such a macroscopic object?

just the same as that discussed several times before (cf. Figure 4.5) but we now include a crucial modification. As shown in Figure 7.1, we imagine that the pointer is mounted on a spring so that, after being deflected to one side, it returns to the centre. It follows that if the photon certainly emerges through, say, the H channel (perhaps because the V channel is blocked) the pointer will swing first to the H side, then to the V and so on, and it will swing in the opposite sense if the photon emerges through V. If, however, the photon passes through both channels and if quantum theory is applicable to both the photon and the macroscopic pointer, the pointer will actually follow the sum of both these motions, which in fact cancel each other out leaving it as it was. If we were to perform such an experiment and observe no motion of the pointer, we would have verified the applicability of quantum theory to such large-scale systems. If on the other hand the measurement chain is broken whenever a many-atom measuring apparatus is involved, the pointer would always swing in one direction or the other when a photon passed through and the quantum cancellation would never be observed.

Unfortunately, such an experiment is impossible to perform on a conventional counter or measuring apparatus because the stationary state at O can be reconstructed only if the two oscillations cancel each other out completely. The pointer is a macroscopic object composed of a very large number of atoms, *all* of which would have to move in exactly the opposite way in the two modes of oscillation. As well as the collective motion associated with the measurement, the position of the pointer is subject to random fluctuations due to its interaction with the air it is passing through and to variations in the frictional forces in the pivot. Even if these could be eliminated, the atoms in the pointer would still possess their more or less random thermal motion, and as a result there is no real possibility of observing a cancellation of the two oscillations. Indeed more detailed calculations show that because of this the pointer would always in practice be found in one oscillation state or the other even if pure quantum behaviour were possible in principle. This has led some scientists to suggest that the presence of such random thermal motion is an essential part of the measuring process and we discuss these ideas further in the next chapter. For the moment however we are interested in the question of whether pure quantum behaviour of a macroscopic body (such as was suggested for the pendulum in the above example) can ever be observed in principle. This question has led a number of scientists in recent years to investigate systems where such phenomena might be detectable and we devote the rest of this chapter to a discussion of this work.

Superconductivity

There is one class of phenomena that is usually thought to exemplify macroscopic quantum effects in a very dramatic way. These are the properties known as superconductivity and superfluidity, both of which are associated with the behaviour of matter at very low temperatures. A superconductor presents no resistance to the flow of electric current so that if such a current is set up in a loop of superconducting wire it will continue to flow for ever, even though there is no battery or other source of electrical potential to drive it. Similarly a superfluid is a liquid that can flow along tubes or over surfaces without friction. Many metals exhibit superconductivity if cooled to a low enough temperature (typically a few degrees above absolute zero) but the only known superfluids are the two isotopes of helium: ^{4}He which becomes superfluid below about 2 kelvins and ^{3}He which exhibits superfluidity only at temperatures less than 0.1 kelvins. The details of the quantum theory of superconductivity and superfluidity are very complex, but they both depend on the quantum properties of a macroscopically large number of particles behaving in a collective manner. The next paragraph describes how this comes about in a superconductor and the relevance of these ideas to the quantum theory of measurement.

The key to understanding superconductivity lies in the fact that at very low temperatures the usual electrostatic repulsion between electrons of similar charge is overcome by an attractive force. (The origin of this attractive force lies in distortions caused in the arrangement of the charged ions in the solid by the presence of the free electrons, but it is not necessary to understand this for our present purposes.) The effect of this attractive force is to create a quantum state of low energy into which the electrons 'condense' – Figure 7.2. They can stay in this low-energy state indefinitely provided they stay there together and an electron can be removed from this collective state only if it is given some energy. In a normal metal, electrical resistance arises because electrons carrying an electric current collide with obstacles (such as impurities or defects in the crystals) and bounce off in a random direction, but in a superconductor such collisions can occur only if the electron gains enough energy to remove it from the condensed state. As the only source of such energy is thermal, it follows that if the temperature is low enough sufficient energy is not available and the electrons have to remain in the condensed state. As a result, the electric current flows without experiencing any resistance and we have an example of a macroscopic quantum effect in which 10^{20} or so electrons all occupy the

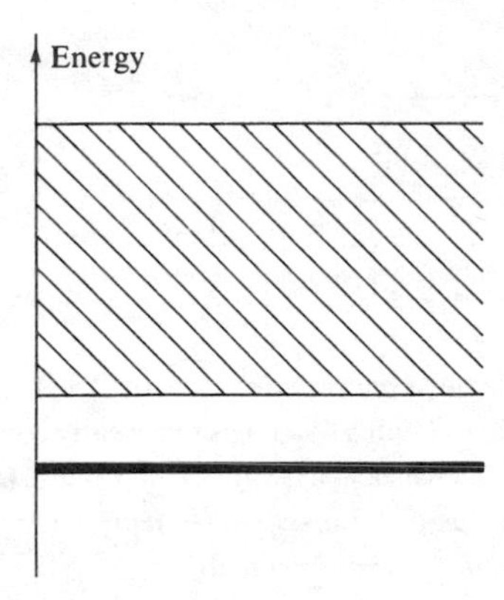

Fig. 7.2 In a normal metal the electrons occupy a broad energy band, but in a superconductor an attractive force between the electrons causes some of them to condense into a low-energy state that is separated from the normal band by a gap.

same quantum state spanning the whole piece of superconductor which is typically several centimetres in size. The experimental observation of superconducting properties seems therefore to confirm the application of quantum physics to such macroscopic systems and might appear to close this way out of the quantum measurement problem.

On closer examination, however, it appears that the many-particle loophole is not quite closed. Although the properties of superconductors confirm the existence of quantum states in which many particles behave coherently, to be relevant to the measurement problem we would have to demonstrate that two such states could be combined coherently in a manner analogous to that suggested for the two pointer oscillations in the previous section. This point was made a few years ago by the theoretical physicist A. J. Leggett who suggested that it might be possible to test such a possibility using a particular kind of superconducting device. His proposal has been the subject of considerable attention by a number of theoretical and experimental research groups in this field and the next section explains the principles behind this project.

SQUIDS

We consider a device known as a SQUID (Superconducting Quantum Interference Device). This consists of a piece of superconductor with two holes through it which very nearly touch each other at a point

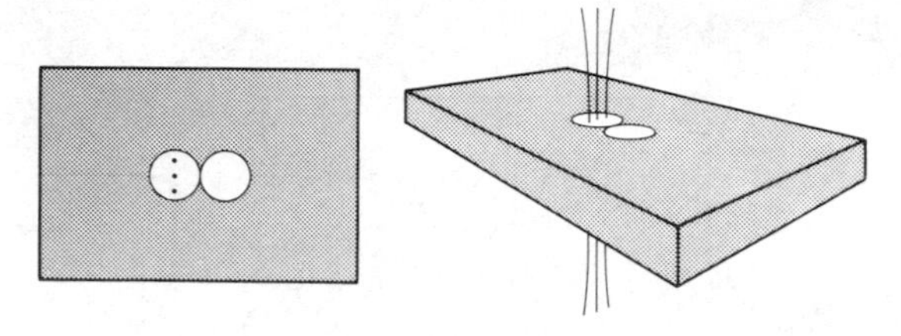

Fig. 7.3 A Superconducting Quantum Interference Device (SQUID) consists of a block of superconductor containing two holes that nearly connect at the 'weak link'. Magnetic flux lines are shown localised in the left-hand hole. Quantum physics predicts that a magnetic flux quantum could be delocalized between the two holes, but this has yet to be confirmed experimentally.

known as the 'weak link' (Figure 7.3). We first imagine the device to be in a state where a current is flowing round one of the holes. As a result a magnetic field is set up whose field lines pass through the hole in the same way as happens in the case of an ordinary electromagnet. One important result of a detailed analysis of the quantum physics of such superconducting systems is that the amount of magnetic flux in such a ring is always a whole number of flux quanta, where the 'flux quantum' has the magnitude $h/2e$ (h being the quantum constant and e the electronic charge). Suppose now that the circulating current, and hence the magnetic field, are very small so that there is only one flux quantum in the left-hand hole. A macroscopically distinct state would be the similar case where one flux quantum threads the other (right-hand) hole. The crucial question is then: can the SQUID exist in a state where the flux quantum is in neither the left-hand nor the right-hand hole, but is delocalized between the two? Quantum physics would say that it can, in the same way as the two oscillation states of the pointer in Figure 7.1 could be combined to cancel out. On the other hand, if quantum theory breaks down for macroscopic objects in such a way as to resolve the measuring problem such a superposition would be impossible. Indeed we can pursue the analogy a little further and imagine the SQUID to be used as part of a device to measure photon polarization: a flux quantum could be created in one hole or the other if the photon were to emerge in the H or V channel respectively of a polarizer. If, on the other hand, the photon had 45° polarization and passed through both channels at once, the SQUID would be put into, or left in, a delocalized state. If this could be demonstrated it would be clear that the measurement chain was not broken at the point of interaction with this macroscopic apparatus. Moreover, because the electrons in the condensed state of a superconductor are not moving thermally, such a measurement might well be possible in practice as well as in theory. We should then be making a direct test of the applicability of quantum theory to a macroscopic measuring apparatus and testing the possibility that it is the number of component particles that distinguishes it from a pure quantum system.

What then is the evidence for or against the delocalization of magnetic flux in SQUIDS? Unfortunately the answer is still far from clear as definitive experiments are very difficult to do. The problem lies largely in the nature of the weak link. Clearly if it were not there at all, we would have only one hole and the flux quantum could be anywhere in it. On the other hand if the barrier is too thick the flux quantum will become effectively trapped on one side or the other and delocalization is not expected. Setting up an experiment with a link of just the right strength to provide a reasonable probability of detectable delocalization turns out to be very difficult. In addition, although the electrons in a superconductor are condensed into a single state, this is not completely insulated from the effects of random thermal motion and the experiments have to be performed at extremely low temperatures. To date, there is some evidence that flux quanta can pass from one hole to another in the SQUID, by what are known as quantum tunnelling processes, but no definitive test has yet been made of the possibility of flux delocalization. Until more experiments are done, all we can say is that there is no evidence as yet for a breakdown of quantum physics in this regime.

There is however a particular reason why a breakdown of quantum theory in the macroscopic situation is very unlikely. This is because unless it happens in a very peculiar way, it opens the door to the possible use of *macroscopic* objects to make measurements on *microscopic* objects that are more precise than quantum physics allows. This is perhaps best understood if we refer to a thought experiment initially discussed by Bohr and Einstein as part of their long debate on the nature of quantum theory (see Chapter 4). They discussed a form of the standard two-slit interference described in Chapter 1 (Figure 1.2) in which the initial slit is mounted on some extremely light springs as shown in Figure 7.4. The purpose of this first slit is to define the starting point of the light beam and it is important to note that, if it is too large, the path difference between light waves passing through different parts of it can be greater than a wavelength and as a result the two-slit interference pattern is washed out. Einstein suggested that if the first slit

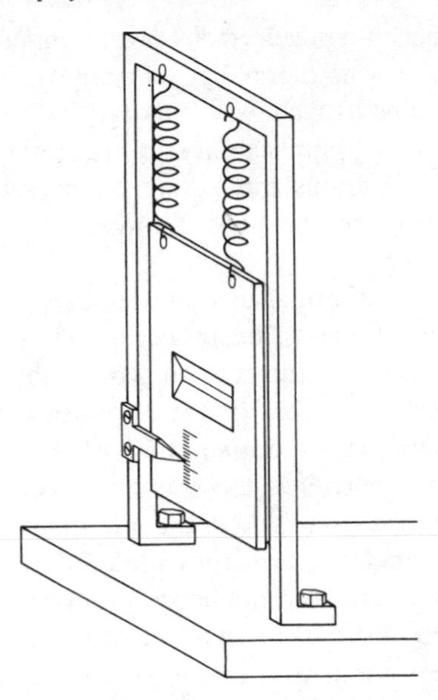

Fig. 7.4 One of the diagrams used by Einstein and Bohr to illustrate their debate about the meaning of quantum physics. If such a device were used as the initial slit of a two-slit interference experiment (cf. Figure 1.2) then the path taken by a photon could in principle be deduced from the direction of recoil of the slit as indicated by the pointer. Bohr showed that provided the slit obeys the laws of quantum physics, such a measurement would destroy the interference pattern.

were to deflect a photon towards the upper of the two slits in the second screen, then the first screen would recoil downwards, while the opposite would happen if a photon were deflected downwards. It would therefore be possible in principle to tell which slit the photon would subsequently pass through by measuring this recoil. However, Bohr pointed out that any such measurement would be tantamount to a measurement of the speed and hence momentum of the recoiling screen and, if it were a quantum object, the Heisenberg uncertainty principle (see page 9) would require a corresponding uncertainty in the screen position. He was able to show that the effective broadening of the first slit implied by this quantum delocalization must be at least large enough to destroy the interference pattern. Wave–particle duality is therefore preserved as is the impossibility of incompatible measurements on photons. It should be emphasized that this experiment was and is a 'thought' experiment that would be quite impossible in practice, but it is conceivable that the principles involved could be tested by experiments on the delocalization of flux in SQUIDS.

If, however, macroscopic objects like SQUIDS or screens carrying slits are not subject to quantum physics and if they really have precisely measurable position and momentum at all times, Bohr's argument no longer holds. We could then extend the principle to the simultaneous measurement of other incompatible properties such as HV and ±45° photon polarizations (e.g. by observing the recoil of the first calcite crystal in Figure 4.2). Similar 'non-interfering' measurements could be made on the photon pairs in an Aspect experiment which would then have to be subject to Bell's Theorem and therefore different from the results obtained by conventional means. As soon as one crack of this kind appeared in the quantum measurement scheme the whole edifice would be in danger of collapse, and it is hard to see how a logically self-consistent theory that would resolve the measurement problem in the macroscopic situation while avoiding this problem could be developed. However, no doubt if experiments on SQUIDS or other systems were to show that this is what happens, such a theory would indeed be devised. Unless and until this experimental evidence is forthcoming, most scientists will believe that such a resolution of the measurement problem is very unlikely.

8 · Backwards and forwards

We saw in the last chapter that direct tests of the pure quantum behaviour of a macroscopic measuring apparatus are very difficult. Because of random thermal disturbances the large number of atoms constituting the oscillating pointer of Figure 7.1 is almost certainly going to swing in one direction or the other, and even the delocalization of magnetic flux in a SQUID has yet to be observed. This leads us to consider another possible way out of the measurement problem. Might it be that random thermal processes are an essential part of any measurement – not just in theory, but in principle? If we could consistently adopt this point of view we could then make the distinction between pure quantum processes and measurements, without directly referring to the size or the number of particles in a system. Thus an ideal oscillating pointer (if such a thing could exist) or a delocalized flux quantum in a SQUID would be in the same category as the 45° photon passing through the HV apparatus: none of these systems could be considered to be in a definite quantum state until this had been measured by an apparatus in which the presence of random thermal motion plays a crucial role.

This approach to the measurement problem will be the subject of the present chapter and the next. We shall see that, although it provides a possible way out of our difficulties, when followed to its logical conclusion it implies a revolutionary change in our thinking about the physical universe that is comparable in its radicalism with any of those previously encountered in the history of scientific thought.

Indelible records

An important feature of any real measurement is that a record is made. By this we do not necessarily mean that the result is written down in a notebook or even that the apparatus is connected to a computer with a memory (although either of these processes would constitute the making of a record) but simply that something somewhere in the universe will always be different because the result of the measurement has some particular value. Thus the photon passing through a detector causes a current to pass which may cause a light to flash or a counter to click to a new position. But even if it doesn't, the current passing through the detector will have caused some slight heating of the air around it which will have moved a little differently than it would have otherwise. Any such change will leave its mark, or record, even if it is so small as to be practically undetectable.

Conversely a quantum event without a measurement leaves no record of its occurrence at all. When the 45° polarized photon passes through the HV polarizer, but then has its original state reconstructed, there is, almost by definition, no record of which HV channel it passed through. In the presence of a detector, on the other hand, the original 45° state can be reconstructed only if, at the same time, we 'erase the record' by returning the counter and everything it has interacted with to its original state. To resolve the measurement problem we need only postulate that such a process is impossible and that the records created by measuring instruments cannot be erased, but are *indelible*. In some ways this does not seem to be a particularly radical postulate – after all we have seen that the probability of all the particles in the detector and its surroundings returning to exactly their original positions is extremely small – but in fact we are making a profound distinction between quantum processes which preserve their full potentiality for reconstructing the original state, and measurement processes in which a record is made and this potentiality no longer exists: the resolution of the measurement problem requires that such records are indelible *in principle*, not just in practice.

Irreversibility

We saw in our discussion of the swinging pointer in the last chapter that it was the random thermal motion of the atoms that in practice prevented the two pendulum motions cancelling out, and also that similar processes make the observation of delocalized flux in a SQUID very difficult to observe. This might suggest that we should look to the branch of physics dealing with such thermal phenomena for a possible understanding of the distinction between processes in which records are made and those which can be completely reversed so that no trace of their occurrence remains. The branch of physics that discusses such phenomena is known as *thermodynamics* and it does indeed contain just such a distinction between what are known as *reversible* and *irreversible* processes. This is contained in what is known as 'The Second Law of

Thermodynamics', a law which had more publicity than most laws of physics ever achieve when C. P. Snow in his famous essay on 'The two cultures' suggested that some understanding of it should be expected of any non-scientist claiming a reasonable knowledge of scientific culture.

There are a number of different ways of stating the Second Law ranging from 'The entropy of the universe always increases' to 'Nothing comes for nothing', but for our purposes the basic idea is probably best summarized by the statement that 'any isolated system always tends to a state of greater disorder'. As an example consider what happens if we have some gas in a container with a partition in the middle (Figure 8.1). Initially we suppose that all the gas is on one side of the partition (this could be achieved by, for example, pumping all the gas out of the other side). If we now open the partition, what will happen? Obviously all the gas will quickly spread out so as to fill the container uniformly and, if we leave it alone, we would never again expect to find it all on the one side. Because the molecules have more room to move around in the whole container than in one section of it, they are more disordered in the former case which is therefore favoured by the Second Law. Another example is the melting of a piece of ice when placed in warm water: the molecules can move more freely and thereby become more disordered in the liquid than when trapped in the highly ordered structure of the ice crystal. This change is therefore allowed by the Second Law while the converse process of a piece of ice appearing in warm water is forbidden and, of course, never observed.

Because they apparently happen in one direction only, processes such as those described are known as *irreversible* processes. Reversible processes, on the other hand, are those that can happen in either direction. Everyday examples of apparently reversible processes are the swinging of a clock pendulum or the spinning of a wheel. Reversible

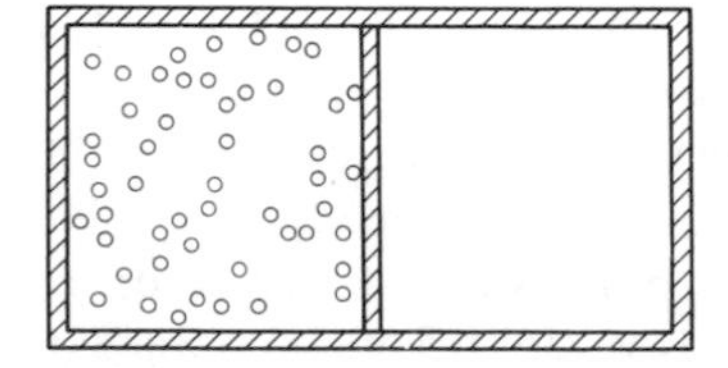
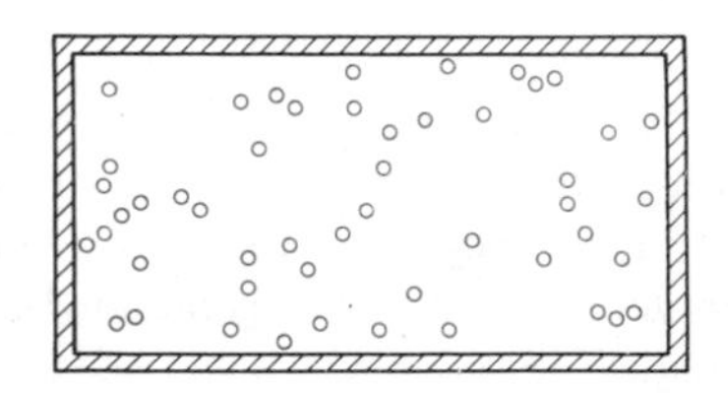

Fig. 8.1 If the barrier confining a gas to the left-hand half of a container is removed, the gas will spread out to fill the whole vessel. The reverse process in which the gas spontaneously moves into one half of the vessel is very improbable and never observed.

and irreversible processes can usually be distinguished from each other by imagining a film to be made of the process which is then run backwards. If the events still appear physically reasonable they are reversible, while if they don't they are irreversible. An educational film produced along these lines shows a ball being thrown to and fro between two people (reversible) and a piece of paper being torn (irreversible).

There are two important ideas in this area that at first sight seem contradictory. The first is that all macroscopic processes are actually irreversible, even if they may appear reversible, and the second is that all the underlying microscopic processes are actually completely reversible! We shall now discuss each of these statements in turn and then the resolution of the apparent paradox.

Why do we say that all macroscopic processes are really irreversible? The crucial idea behind this statement is that it is never possible to isolate a macroscopic body completely from its surroundings. Thus, although the swinging pendulum appears to be behaving reversibly, careful examination will show that its motion is actually slowing down, albeit extremely gradually, due to drag forces from the surrounding air or friction at the pivot bearings. Alternatively, if, like a clock pendulum, it is driven by some kind of motor, examination of this power source will reveal irreversible processes such as a falling weight, an unwinding spring or a discharging battery. The most reversible large-scale process we could think of might be a moon orbiting a planet or a planet orbiting the sun. These objects are passing through the vacuum of space where there is nothing to exert drag forces and their motion has been going on for millions of years without stopping. Nevertheless it is known that there are frictional forces even in this situation: for example the tides on the earth caused by the moon and the sun dissipate energy which comes from the orbital motions which in turn are slowed down. As a result the average distance between the earth and moon increases by a few millimetres each year, an effect that has recently been measured directly by laser-ranging experiments.

We turn now to the second statement in the paradox: that all fundamental microscopic interactions are reversible. This arises because the laws of mechanics and electromagnetism do not depend on the direction (or 'arrow' as it is sometimes called) of time. A simple situation illustrating this is a collision between two molecules, as shown in Figure 8.2, from which we see that such a process is just as acceptable physically if the direction of time is reversed. This is true of all physical events occurring at the microscopic level (apart from one or two processes involving particular subatomic particles which do not need to

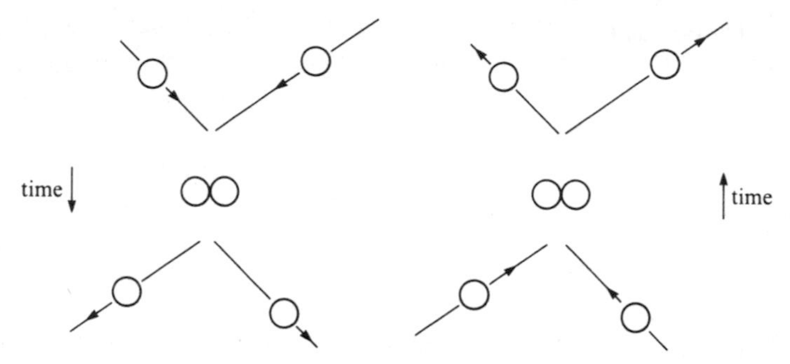

Fig. 8.2 Elementary dynamical processes, such as the collision of two spheres, appear just as physically reasonable if their time order is reversed.

concern us). How then does it turn out that the behaviour of large-scale objects is always irreversible when the motion of their atomic constituents follows these reversible macroscopic laws?

The standard answer to the problem of macroscopic irreversibility arising from microscopic reversibility is that the irreversibility is an approximation or even an illusion. Consider again the case of the container of gas (Figure 8.1) whose partition is removed and the gas then fills the whole vessel. If we were to concentrate on a single gas molecule we would find it undergoing a series of collisions with the walls of the vessel and with other molecules, each of which is governed by reversible laws. Suppose that after a short time we were able to reverse the direction of motion of every molecule in the gas. Because the mechanical laws governing the collision work just as well 'in reverse', each molecule would retrace the path it had followed and after a further time identical to that between the opening of the partition and the reversal the molecules would all be back in the left-hand part of the container and we could close the partition and so restore the original state of the system. An example of this process being followed by a 'gas' of three molecules is illustrated in Figure 8.3. Of course stopping and

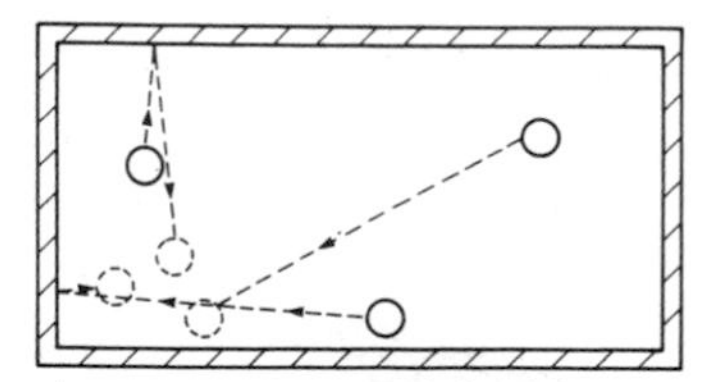

Fig. 8.3 Although a gas is never seen to move into one half of its container, this is not forbidden by the laws governing atomic collisions. The diagram shows three particles initially scattered around the container (solid circles) moving in such a way as to bring them together in the bottom left-hand corner (broken circles).

reversing the molecular motion is not possible in practice, but it is apparently always possible in principle. Moreover, a simple analysis of the motion leads us to believe that, even in the absence of intervention, a physical system will always return to its initial state if we wait long enough.

To understand this last point, we first consider the very simple example of a single molecule moving inside a flat rectangular area and bouncing off its boundaries as in Figure 8.4. If it is set off moving parallel to one of the sides of the rectangle, then it will bounce up and down a line indefinitely. Every other time it passes the starting point, it has the same speed in the same direction as it had originally, so in this case the recurrence happens very quickly. Similarly if the molecule starts off from a point on one side at an angle of exactly 45°, it will trace out the path shown in Figure 8.4(b) and continue to follow this course indefinitely. Other simple repeating paths can easily be devised. However, in the case of a more general starting angle the path will be very complex and an imaginary line drawn by the particle will eventually fill the whole square. It is easily seen that the simple repeat occurs only if the tangent of the starting angle is a rational number – that is if it is equal to n/m where n and m are two whole numbers. This is infinitely less probable than the irrational case, so, in general, we can assume that, if

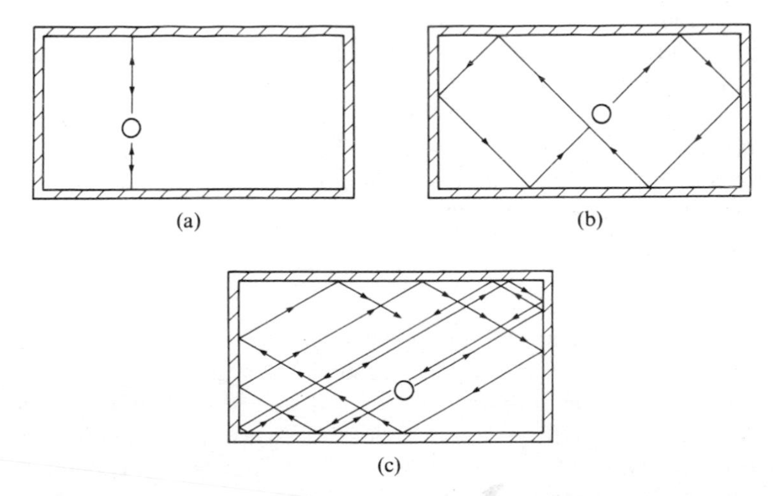

Fig. 8.4 A single particle moving in a rectangular container can undergo different kinds of motion depending on its starting conditions. In (a) and (b) the motion repeats cyclically after a few collisions. Far more likely is a path such as that shown in (c) which never returns to its precise starting condition. In this latter case the ergodic principle implies that the particle will eventually occupy all possible dynamical states.

we wait long enough, the particle will have occupied every possible position on the square. (The reader who has access to a microcomputer may wish to study this behaviour by programming it to display a point moving in these ways on the screen.) It follows that such a system, if left to itself, will sooner or later pass through every state that is consistent with the principle of conservation of energy. This last statement is known as the *ergodic hypothesis* and is generally assumed to be true for any physical system such as a gas of molecules. Of course in the latter case a description of the physical state of the system requires a knowledge of the position and velocity of each of the 10^{22} or so molecules in the gas, so the number of possible states to be passed through is immense. If the ergodic hypothesis is correct, however, all possible states will eventually occur and some of these states, even if they are not precisely identical to the initial state of the gas will be arbitrarily close to it. It follows therefore that the gas must return to a state that is arbitrarily close to its initial state if we wait long enough. This fact was first realized in the nineteenth century by the French physicist Henri Poincaré: the return of a thermodynamic system to an earlier state is known as a 'Poincaré recurrence' and the time it takes for this to happen is known as the 'Poincaré cycle time'.

It is arguments such as these that lead to the belief that irreversibility is an illusion. Although the gas expands to fill the empty box, there is always some chance, however small, that it will return to its initial state with all the molecules in the left-hand half of the container, and this will certainly happen if we wait long enough. Of course we may have to wait a very long time. Estimates of Poincaré cycle times for typical macroscopic systems come out at many millions of times the age of our universe, but there is always a possibility, however remote, that the recurrence will occur at any time.

Irreversibility and measurement

It should now be clear why any appeal to thermodynamic irreversibility does not immediately resolve the measurement problem. It is certainly true that pure quantum processes are strictly reversible in a thermodynamic sense while measurements always involve thermodynamic irreversibility; but if the ergodic hypothesis and the idea of the Poincaré recurrence are correct this irreversibility is an illusion arising from the fact that we cannot observe any large-scale process for a long enough time to expect to see a Poincaré recurrence. If we could make such an observation, the atoms in the swinging pointer *would* eventually happen to be in exactly opposite positions when swinging in the two directions and as a result the two motions would cancel out, and the polarization state of the photon would still be 45° and neither H nor V. Even if some other record had been made of which way the pointer was swinging, this must also be constructed out of materials that are subject to the laws of thermodynamics and so the whole lot together can be considered as a single thermodynamic system and subject to an eventual Poincaré recurrence. If there is no point at which we can say that the measurement has finally been made then the Copenhagen interpretation insists that we must not ascribe any reality to the unmeasured quantities (the photon polarization, the direction of the pointer swing or indeed anything else!) and we are still impaled on the horns of the measurement problem.

If we are going to make progress in this direction we are going to have to find some reason why the ergodic hypothesis and the ideas of Poincaré are not applicable to a quantum measurement. One suggestion that has been made is to note that no real thermodynamic system is ever completely isolated from its surroundings. Thus the atoms in a gas interact with external bodies, if only through the weak gravitational interactions that operate between all objects. Although such influences are extremely small, it has been shown that very tiny perturbations can have a large effect on the detailed motion of a complex many-particle system like a gas or, indeed, a pointer: thus a small change in a distant galaxy could affect the behaviour of a container of gas on the earth to such an extent as to make a major change in the time of a Poincaré recurrence. It follows that no complex system can be considered as having returned to a state it was in earlier unless its whole environment is also the same as it was previously – because, if this is not the case, its future evolution will not be the same as it was the first time round. We could therefore suggest that a measurement represents an irreversible change in the whole universe and simply postulate that the laws of quantum physics are applicable to any part of the universe, but not to the universe as a whole. We might even argue that it is meaningless to consider the possibility of the universe returning to a state it had occupied previously as this would imply that the state of all observers in the universe would have to be similarly restored and there is no way we could know that anything had occurred or that time had passed. There are two objections to this line of thinking. First, incredible as it may seem, it does appear as if quantum theory can be applied to the universe as a whole. Recent theoretical studies of its development during the very early stages of the 'big bang' rely on just this postulate and produce results that appear to explain some of the observed properties of the universe today. Secondly, it is difficult to decide when an irreversible change in the whole universe has occurred. All physical influences including gravity travel at the speed of light or slower so it would take many years before the change associated with a measurement was registered in the distant parts of the universe. Clearly we cannot refuse to ascribe reality to an event for the first few thousand million years after its occurrence!

There is, however, a completely different approach to the problem and this is simply to postulate that the ergodic hypothesis is not correct and that Poincaré recurrences do not happen. After all, no breach of the Second Law of Thermodynamics has ever been observed in practice and the whole argument rests on some really quite sweeping hypotheses. Could we not find some aspect of the nature of the thermodynamic changes that occur in measuring processes which would clearly distinguish them from the kind of process for which reversibility is a possibility and to which pure quantum theory can be applied? This idea has been suggested on a number of occasions, and it has been developed in recent years by Ilya Prigogine who won a Nobel prize in 1977 for his theoretical work in the field of irreversible chemical thermodynamics. His ideas will be discussed in the next chapter where we shall see that they suggest a possible solution to the measurement problem that is more down to earth than any we have discussed so far, but which still involves a revolutionary change to the conventional view of the physical universe.

9 · Only one way forward?

The starting point of the approach by Prigogine and his co-workers is a re-examination of the validity of the ergodic principle which leads to the idea of the Poincaré recurrence. We made the point in the last chapter that, in the simple case of a single particle confined to a rectangle, if the starting angle does not have a special value the particle trajectory will fill the whole square and the particle will sooner or later return to a state that is arbitrarily close to its initial state. By this we mean that, although the initial and final states cannot be precisely identical, we can make the difference between the initial and final positions and velocities as small as we like by waiting long enough. The implicit assumption is that the future behaviour of the system will not be significantly affected by this arbitrarily small change in state: its behaviour after the recurrence is expected to be practically the same as its behaviour was in the first place and if it is part of a quantum system the recurrence is said to reconstruct the original quantum state so precisely that we cannot say that any change really took place. This idea is certainly applicable to a simple system like a single particle bouncing around a square, but it turns out that it has only limited validity in the case of more complex systems.

As an example of a physical system whose future behaviour can change drastically as a result of an arbitrarily small change in its state, consider a simple pendulum constructed as a weight at the end of a rod which is mounted on a bearing so that it can rotate all the way round (Figure 9.1). This has two possible quite distinct types of motion. If it is pulled a bit to one side and released, it will swing backwards and forwards like a clock pendulum, but if it is given a large initial velocity it will swing right over the top and continue rotating until it slows down. At any point in the motion we could nearly always tell by measuring its speed whether it has enough energy to rotate or not. However, the important point is that there is a critical speed below which the pendulum will just be able to swing up to the highest point and then will swing back, and above which rotation will occur. It follows that the future behaviour of two such pendulums whose only difference is the fact that their velocities are higher and lower than this critical velocity

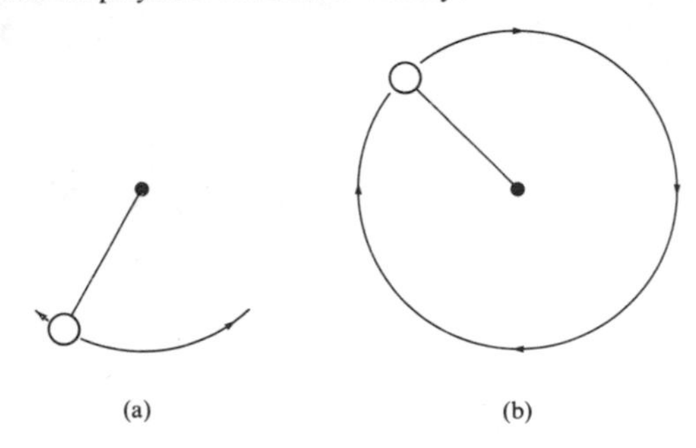

Fig. 9.1 A simple pendulum can undergo two distinct types of motion, either oscillation as in (a) or full rotation as in (b).

by arbitrarily small amounts will be completely different. Consider now what would happen if such a pendulum were part of a thermodynamic system. For example we might imagine a very small version of it in a container along with a number of gas molecules so that from time to time the molecules strike the pendulum giving it various amounts of energy. Suppose now that at some point the pendulum has very nearly, but not quite, enough energy to start rotating; it could well be that at some later time the system could return to the same state, with the molecules in practically the same place moving with almost the same speed, but with the energy given to the pendulum just slightly greater than before. The pendulum would now start rotating so that its whole future behaviour and that of the molecules colliding with it would be quite different from that which followed the initial state. It should now be clear that there are difficulties in applying the ergodic principle and the idea of a Poincaré recurrence to a system like this: however similar the initial and final states of such a system may be, their future behaviour may be completely different.

It might seem that a situation like the above is so unusual as to be practically irrelevant. However, the details of the motion in many-particle systems are extremely complex and such instabilities are actually quite commonplace. So much so in fact that many systems are subject to even stronger instabilities such that *any* arbitrarily small change in the starting conditions drastically alters its future behaviour. The effect of such instabilities on a physical system is sometimes known as 'strong mixing' and systems that are liable to this behaviour are said to have the property of 'weak stability'. It has been shown that a 'gas' composed of as few as three hard-sphere particles behaves in this way so it is very likely that strong mixing is a very common feature of most if not all real thermodynamic systems.

Following Prigogine, we first consider the relevance of strong mixing to our understanding of classical physics and return to the quantum case and the measurement problem a little later. The first point to note is that, even although the detailed future behaviour of the component particles of a system is unpredictable in the presence of weak stability, this does not mean that none of its physical properties are measurable. In particular, the traditional thermodynamic quantities such as the temperature and pressure (if it is a gas) are perfectly well defined. Moreover, the Second Law of Thermodynamics is strictly obeyed, the system tending to a state of greater disorder even more rapidly than it would if subject to the ergodic principle. It is the underlying dynamical parameters, such as the positions and velocities of the component molecules, that are changing chaotically, whereas the thermodynamic quantities that are traditionally thought to be derived from the microscopic substructure are well behaved. These facts led Prigogine to suggest that our traditional way of thinking about thermodynamic systems is wrong and that, if strong mixing is present, we should consider the thermodynamic quantities to be the primary reality and the allegedly more fundamental description in terms of microscopic structure to be invalid. He was able to show that it would then be possible to treat the Second Law of Thermodynamics in the way its name suggests: as a *law* that is always followed rather than a statistical rule that would always be eventually broken by a Poincaré recurrence.

The consequences of this way of thinking are profound even before we consider the extension of the ideas to quantum theory. If we follow Prigogine's approach, indeterminism becomes an implicit part of *classical* physics. Our inability to predict the future motion of the components of a many-particle system is no longer to be thought of as a limitation on our experimental or computational ability, but as an inevitable consequence of the laws of nature. Instead of these fundamental laws referring to microscopic reversible processes, with macroscopic irreversible behaviour as an approximation or illusion, it is the irreversible laws that should be taken as fundamental and reversibility which is the approximation. Prigogine postulates a kind of uncertainty principle linking the two types of description: if the thermodynamic description is appropriate, precise measurement of the dynamical variables is impossible, while a simple system (like the single particle on the square) that can be described dynamically possesses no definite thermodynamic parameters. We might even go so far as to extend the ideas of the Copenhagen interpretation to this field and suggest that it is just as meaningless to ascribe reality to the positions and velocities of the individual particles in a system subject to strong mixing as it is to talk about the temperature of a single isolated particle. It is interesting to observe some of the ideas of quantum physics showing up in our thinking about this completely classical system.

The idea that it is reversibility that is the approximation to irreversibility rather than the other way round is certainly more consistent with our view of the universe as we see it. Thus (again ignoring quantum ideas to which we shall be returning very soon) all real systems are visibly subject to the laws of thermodynamics and are 'running down'. Even if we wait a long time we don't actually expect to see a Poincaré recurrence and anyone who claimed to observe one in a system subject to strong mixing would be disbelieved as much as any other reporter of an alleged miracle. If the Second Law is fundamental, then this running down is predicted and the Poincaré recurrence is not expected. The time to run down varies very much from system to system. Thus the block of ice in warm water melts in a few minutes, but it will be many millions of years before the solar system is consumed by the sun, even longer before the galaxy collapses and longer still before the universe eventually comes to an end. Our ability to predict the future behaviour of physical systems seems to be greater for the large-scale parameters of large systems. Thus we are unable to predict the future trajectory of a molecule in a gas more than a few collisions ahead, but we can work out to considerable precision how the planets will be moving round the sun in some thousands or even millions of years' time. Eventually, however, the behaviour of the solar system becomes unpredictable and dominated by the strong-mixing interactions between its components. In the very long term the only things we can predict are the global parameters of the universe (its overall size, mass, temperature, etc.) and the fact that these cannot be accurately predicted at the present time is probably due to the inaccuracy of our data rather than fundamental limitations of the type we have been considering.

Return to the measurement problem

Everything we have said so far in our discussion of strong mixing has referred to classical systems, but it is now time to return to the main point at issue, the measurement problem. Superficially at least the solution is very simple. If we postulate that the Second Law is fundamental for all physical systems that are subject to strong mixing and that a Poincaré recurrence can never occur in such cases, then it is a short step to say that all quantum measurements are carried out with apparatus of this kind and the measurement chain is broken whenever strong mixing becomes involved. This is indeed Prigogine's solution to the measurement problem, but he also goes further than this and explores some of the consequences this picture has on our view of the quantum world.

In the same way as it was always believed that the fundamental parameters of a classical system are the positions and velocities of its components and that the thermodynamic description is a statistical approximation, the traditional approach to quantum physics has always been to give primacy to the pure quantum states of the system and to look on the measurement interactions as secondary approximations. Thus, as we have seen, the whole measurement problem arises when we treat the measurement apparatus as a pure quantum system. Prigogine's approach is to reverse this way of looking at things: the behaviour associated with a quantum measurement is now taken as the fundamental reality with pure quantum behaviour as an approximation appropriate only to the special situations where the effects of strong mixing, always present in a measurement, can be neglected.

This new approach also changes the way we look at the fundamental indeterminacy traditionally associated with quantum physics. We note first that the behaviour of a pure quantum system in the absence of a measurement is actually quite predictable: the 45° photon passing through the HV polarizer does not pass through one channel or the other, but in some way we find very difficult to model it passes through both. If it didn't we wouldn't be able to reconstruct the original state in the way we have so often discussed and, even in the absence of such a reconstruction, quantum theory describes the photon as being a well defined state that is a superposition of an H photon in one channel and a V photon in the other. The indeterminacy arises when we allow it to interact with a measuring apparatus which results (*pace* many-world supporters) in the state collapsing at random into one polarization or the other. But we have seen earlier that there is a fundamental indeterminism, quite independent of quantum effects, associated with the strong mixing always implicit in a measurement. Is there then any need to introduce an additional uncertainty associated with the quantum system? The answer is no. Prigogine has shown that the extension of his ideas to the quantum regime actually introduces extra correlations that reduce the indeterminism associated with strong mixing in the classical situation, to a point consistent with the quantum uncertainty principle.

The distinction between irreversible measurements and reversible pure quantum processes makes it a little easier to accept some of the peculiar features of the latter. If there is no irreversible change associated with a quantum event, it is perhaps hardly surprising that we find it difficult to describe such processes in conventional terms. After all, if we knew through which HV channel the photon had passed, an irreversible change would have occurred somewhere, if only in our brains, and the process would no longer be completely reversible. A pure quantum process occurs only in a parameter, or set of parameters, that have become detached from the rest of the universe, and perhaps even from space-time itself, and leave no trace of their behaviour on the rest of the universe until a measurement interaction takes place. We should perhaps be more surprised by the fact that quantum theory allows us to say anything at all about the behaviour of a quantum system between measurements, than by our inability to make a precise description in this realm.

We see now why Prigogine's theories, although apparently simple, involve a quite revolutionary change in our thinking about the physical universe. For a long time now the emphasis in physics has been on measuring and understanding the behaviour of the elementary, subatomic particles that are believed to be the fundamental building blocks of nature. It is implicitly, or sometimes explicitly, stated that the behaviour of macroscopic bodies, or even the universe at large, can be understood purely in terms of these elementary particles and the interactions between them. Thus the behaviour of a gas composed of such particles, all moving subject to reversible laws, must also be reversible, and the apparently irreversible changes must be an approximation or illusion resulting from our observation over too short a time scale. Prigogine completely reverses this way of looking at things. He suggests that it is the irreversible changes that are the really fundamental entities in the universe and that the idea of microscopic particles moving subject to reversible laws is an approximation that is valid only in the very special circumstances where a particle, or particles in cooperation, are effectively decoupled from their interaction with the rest of the universe. Notice that the fundamental concepts are the events or changes rather than the objects that are doing the changing. Prigogine's own words to describe this change of emphasis from 'being' to 'becoming' are:

The classical order was: particles first, the second law* later – being before becoming! It is possible that this is no longer so when we come to the level of elementary particles and that here we must *first* introduce the second law before being able to define the entities. Does this mean becoming before being? Certainly this would be a radical departure from the classical way of thought. But, after all, an elementary particle, contrary to its name, is not an object that is 'given'; we must construct it, and in this construction it is not unlikely that *becoming*, the participation of the particles in the evolution of the physical world, may play an essential role.

Although this change of emphasis may seem revolutionary, it is, as we have seen quite consistent with our experience of the physical universe. Any experience we have is certainly of irreversible processes involving strong mixing, if only because the changes taking place in our brain are of this nature. By definition we have no experience of reversible, pure quantum 'events' that are not detected. But this is not to say that such a theory is in any way subjective in the sense discussed in Chapter 5. It is not the fact that pure quantum 'events' have no effect on *us* that is important, it is that they result in no permanent change to any part of the universe. The laws of classical physics were set up on the unquestioned assumption that, although events may be reversible, it is always possible to talk about what has happened. Even Einstein's theory of relativity refers extensively to the sending of signals which are clearly irreversible measurement-type processes. Perhaps it should not be surprising that, when we try to construct a scenario that goes beyond the realm of possible observation into the reversible regime, our models involve apparent contradictions such as wave–particle duality and the spatial delocalization observed in EPR experiments.

One of the advantages of this way of looking at the measurement situation is that it actually brings us back to something very like the Copenhagen interpretation. Those of us trained in this point of view have just about learned not to attribute 'reality' to unobservable quantities. Thus questions like whether an object was really a wave or a particle, or whether a 45° polarized photon really went through the H or the V channel could never be answered in principle and so there is no point in asking them; the emphasis should always be on the prediction and understanding of the results of measurements made by observers. As we have seen in the last few chapters, the problem arises when we attempt to make a consistent distinction between the observer and the observed object. Prigogine's approach is not to distinguish between

* i.e. the Second Law of Thermodynamics.

these entities, but between the nature of the processes. If we identify the idea of a measurement by an observer in the Copenhagen interpretation with an irreversible change in the universe brought about by the onset of strong mixing, we may well have obtained a consistent solution to the measurement problem.

Despite the successes of the approach outlined in this chapter, the reader should not think that we have now reached a final conclusion accepted as orthodox by the general body of physicists. There are probably two main reasons for this. First, the correctness of the distinction between genuinely irreversible processes involving strong mixing, and reversible, ergodic changes is not universally accepted. It is difficult to give up the idea that if only we knew the initial state exactly enough, we could predict its future behaviour and that its earlier state actually would recur if we only waited long enough. There are still many technical problems in the understanding of irreversible systems and not all workers in the field believe that these ideas satisfactorily solve the measurement problem. Secondly, Prigogine's idea that irreversibility should be postulated as fundamental, with the reversible changes occurring only in particular special circumstances, is too radical an approach for some. The idea of the physical universe as an assemblage of interacting but independently existing microscopic particles is not only deeply rooted in our thinking, but is considered by some to be essential to a complete physical theory. A model of the physical world that attributes all reality to the changes, while stating that it is impossible to make a consistent description of what it is that is changing, is difficult to accept.

A further objection that could be made is that, unlike previous revolutions in scientific thought, Prigogine's theories are not obviously testable by experiment: he simply says that the coherent reconstruction of a quantum event should be postulated as impossible in principle rather than treated as unobtainable in practice. Of course this objection is valid for nearly all the measurement theories we have discussed: although the idea that it is the actual number of component particles that is important (Chapter 7) is subject to a possible direct test which has yet to be carried out, the others do not seem to be distinguishable by any direct experiment, and, to some extent at least, we are therefore free to choose the one most suited to our philosophical prejudices. Those of the author are probably becoming fairly clear by now, but the last chapter of this book brings them out in a more explicit form.

10 · Illusion or reality?

Now that we have completed our survey of the conceptual problems of quantum physics and some of their possible solutions, what are we to make of it all? One thing that should be clear is that there is wide scope for us all to have opinions and there is a disappointing lack of practicable experimental tests to confirm or disprove our ideas. Because of this I intend to drop the use of the conventional scientific 'we' in this chapter and to use the first person singular pronoun wherever I am stating an opinion rather than describing an objective fact or a widely accepted scientific idea. This is not to say that everything in the earlier chapters has been free of personal bias, but I have tried to maintain a greater basis of objectivity there than will be appropriate from now on.

I first want to refer briefly to a way of thinking about philosophical problems known as 'positivism'. Encapsulated in Wittgenstein's phrase 'of what we cannot speak thereof should we be silent', positivism asserts that questions that are incapable of verification are 'non-questions' which it is meaningless to try to answer. Thus the famous, if apocryphal, debate between mediaeval scholars about how many angels can dance on the point of a pin has no content because angels can never be observed or measured so no direct test of any conclusion we might reach about them is ever possible; opinions about such unobservable phenomena are therefore a matter of choice rather than logical necessity. Positivism can often exert a salutory beneficial influence on our thinking, cutting through the tangle of ideas and verbiage we (or at least I) sometimes get into, but if taken too far it can lead us to a position where many obviously acceptable and apparently meaningful statements can be dismissed as meaningless. An example often quoted is a reference to the past such as 'Julius Caesar visited Britain in 55 B.C.', or 'Florence Nightingale nursed the troops in the Crimean War'. There is no way such statements can be directly verified as we cannot go back in time to see them happening, but everybody who knows some history would believe them to be both true and significant and certainly not meaningless. Or what about a statement about the future such as 'the world will continue to exist after my death'? There is no way that I can directly verify this proposition, but I firmly believe it to be both meaningful and true. A positivist analysis can be very useful, but it should be employed with caution.

Another word used to categorize statements in science is 'falsifiability'. The importance of this concept was first stressed by the philosopher Sir Karl Popper (many years before he developed the ideas about the mind discussed in Chapter 4). He suggested that the difference between scientific and other kinds of statements is that it should always be possible to think of an experimental test with a possible outcome that would show the statement to be false. Thus Newton's law of gravitation would be shown to be false if an apple or other object were to be released from rest, but not fall to the ground. Many theories about events outside science, perhaps particularly religion and politics, are not subject to such a test: whatever facts are observed, the tenets of such a theory remain intact. Some would claim that unfalsifiable statements are meaningless, but Popper always argued against this and the positivist viewpoint in general.

A form of positivism is clearly a very important part of the thinking behind the Copenhagen interpretation. If the very nature of the experimental set-up excludes any possibility of the measurement of a physical quantity, can that quantity have any 'reality' or is it in these circumstances just an 'illusion'? Certainly, as we saw in Chapter 2, the idea of a horizontally polarized photon being also polarized at 45° to the horizontal would be a contradiction in terms if applied to the idea of polarization in the wave context, but is it similarly meaningless to ask which slit a photon passed through in the Young's interference experiment? As someone educated in (or perhaps brainwashed by) the Copenhagen tradition, I say 'Yes, this is an illusion, the particle does not have a position – it is not really a particle – unless the experiment is designed to make a measurement of this property'. However, I am very aware that this kind of thinking does not come easily or naturally but seems to be forced on us by the development of quantum physics. In the nineteenth century some thinkers argued from a positivist viewpoint that the idea of matter being composed of atoms was a similarly meaningless postulate which could not be directly tested, but we all now accept the reality of the existence of atoms as a directly verifiable objective fact. Could it be that the Copenhagen interpretation is wrongly encouraging us to classify as illusion quantities that are perfectly real and will be observed when our knowledge and technology progress far enough? It is thinking like this that makes the idea of hidden variables seem both plausible and attractive – if it were not that no hidden-variable theory (certainly none that preserves locality) is capable of predicting the results of two-photon correlation experiments, such as the Aspect experiment discussed in Chapter 3! Although the atomic postulate was not *necessary* to explain the known phenomena in the last century, it was still a perfectly tenable hypothesis. In contrast, Bell and Aspect have shown that a local hidden-variable theory is untenable and incapable of accounting for the observed phenomena and therefore cannot be believed in, however attractive such a belief might be. If things had turned out differently and a successful hidden-variable theory had been developed on the basis of a simple model of the microscopic world, there is no doubt that it would have quickly gained wide acceptance and the traditional quantum theory would have been abandoned – even if both theories had made identical predictions of the results of all possible experiments. Positivists might have said that there was no meaningful distinction between the two approaches, but, just as in the case of the atomic hypothesis, nearly everyone would have preferred the view based on a realistic model of the microscopic world. It is because this has not happened that I, along with most other physicists, have had to accept the Copenhagen ideas – not because we particularly wanted to, but because this is the only way we can come near to describing the behaviour of the physical world. As Bohr pointed out a number of times it is nature itself and not *our* nature which forces us into this new and in many ways uncomfortable way of thinking.

Nevertheless the Copenhagen interpretation leaves us with the measurement problem. If reality is only what is observed and if quantum physics is universal who or what does the observing? If the cat is a quantum object, what is it inside or outside the box that decides whether it is alive or dead or if it even exists? At this point the pure positivist might like us to go no further. 'It is meaningless to ask', she might say, 'whether it is your consciousness, the macroscopic apparatus or the irreversible change that causes the cat to be alive or dead – or indeed if there is a multiplicity of cats each in their own universe – as there is no experiment to decide one way or the other'. Perhaps this is so, perhaps it is a non-question, but is a fascinating non-question about which I have opinions that are real to me at least.

If it is meaningless to distinguish between the idea of an objective real world 'out there' and the proposition that everything is just my sense impressions, then I feel it hard to think of any statement that is meaningful. The aim of science must be to attain an objective description of the physical universe. It may be that a theory based on consciousness and subjectivism could be consistent with the observed facts, but I find its implications, such as the non-existence of a physical universe until a mind evolved (from what?) to observe it, quite unacceptable and I would prefer to believe almost any theory that preserved some form of objectivity.

What about the other extreme view where an unimaginably huge number of other universes 'really exists' and the idea that we have only one self in a single universe is an 'illusion'? The fact that this theory has any credence at all certainly belies the idea that all scientists are positivists! It is difficult to think of a more extreme example of the postulation of unobservable quantities to overcome a scientific problem, and one really does not need to be much of a positivist to reject the idea of many unobservable universes as meaningless and to compare the estimation of their number to the traditional counting of angels. And yet the many worlds have their attractions. These lie largely in the fact that this is the only theory that seems to preserve quantum physics as the single universal theory of the physical world, apparently capable of describing all phenomena from the smallest to the largest without further postulates – 'cheap on assumptions' even if 'expensive on universes'. And yet, as we saw, some at least of the advantages of the many-universe model are illusory rather than real. It overcomes the non-locality of the Aspect experiment only by making it unnoticeable in the much more dramatic non-locality of the branching universe. And if branching occurs only when a measurement-like interaction takes place, the measurement problem remains unsolved and the only gain left to the many-worlds model is the recovery of some form of determinism.

One of the features of some writing on the quantum measurement problem is the statement, or at least implication, that there are only two possible views – subjectivism or many worlds – so that if one is rejected the other must be accepted. Thus in an article written in *Psychology Today* by Harold Morowitz in 1980*, the subjective interpretation of the quantum measurement problem is described as the standard view of physicists and this is contrasted with the mechanistic approach to the mind adopted by many modern biologists. In countering this view, however, the writer Douglas Hofstadter* puts forward the many-worlds interpretation as the only alternative. If these were indeed the only alternatives, I should certainly choose the objective many-worlds rather than the subjectivist viewpoint, but I think we are a long way from

* in *The Mind's I* cited in the bibliography.

having to make such a choice. The seeds of another approach were in fact already present in some of Niels Bohr's writing on the Copenhagen interpretation. Although his frequent references to 'the observer' have misled some writers into thinking that the Copenhagen interpretation is essentially subjective, this is not the case. Bohr was always at pains to emphasize the importance of the disposition of the measuring apparatus rather than any direct influence emanating from the experimenter. Thus

> Every atomic phenomenon is closed in the sense that its observation is based on registrations obtained by means of suitable amplification devices with irreversible functions such as, for example, permanent marks on a photographic plate cased by the penetration of the electrons into the emulsion.
>
> N. Bohr: *Atomic Physics and Human Knowledge*, Wiley, New York (1958)

and

> . . . it is certainly not possible for the observer to influence the events which may appear under the conditions he has arranged. To my mind, there is no alternative than to admit that, in this field of experience, we are dealing with individual phenomena and that our possibilities of handling the measuring instruments allow us only to make a choice between the different complementary types of phenomena we want to study.
>
> In *Albert Einstein: Philosopher-Scientist*, P. A. Schlipp (ed.), pp. 200–41, The Library of Living Philosophers, Evanston (1949)

On the other hand, Bohr also emphasized the importance of applying quantum physics to macroscopic objects, showing that this was necessary to preserve the consistency of quantum theory and to prevent the possibility of making measurements on the atomic scale that were inconsistent with the uncertainty principle and the general laws of quantum physics. However, he does not seem to have seen the potential contradiction between these two approaches to the macroscopic world or to have seriously considered the Schrödinger-cat measuring problem.

Of course it doesn't really matter what Bohr or anyone else said or thought; science unlike law or theology defers not to past authority, but to the way nature is. It should be clear from the later chapters in this book what a huge step is taken in going from the measurement problem to either the subjective or the many-worlds postulate. Our experimental knowledge of the application of quantum physics to the macroscopic world is so limited, and the qualitative difference between, on the one hand, the quantum process of the photon or electron passing through the polarizer or interference slits and, on the other, the recording of the arrival of a particle on the emulsion of the photographic plate is so great that it must surely be important to eliminate the possibility of a solution in this area before moving into more fanciful regimes.

It is therefore of supreme importance that the predictions of quantum physics be tested in the macroscopic regime wherever possible and this is one of the most important byproducts of the Aspect experiment. The investigation of the detailed properties of superconducting devices is certainly a promising area for testing the applicability of quantum theory to phenomena involving the coherent behaviour of a large number of particles, and this is a field to which I devote some of my own research effort. Despite this I believe the likelihood of a falsification of quantum physics in this area to be very small. As pointed out by Bohr and as outlined at the end of Chapter 7, if such coherent macroscopic motion is not subject to quantum theory, it is hard to see how quantum physics can be applied consistently to atomic phenomena either.

This brings us back to the idea of the measurement as a thermodynamically irreversible process. This view is often criticized on the basis that the inevitable, if eventual, Poincaré recurrence means that quantum coherence must be maintained and it is never correct to say that a measurement has finally been made. Nevertheless, there is a practical operational distinction between thermodynamically irreversible changes and the reversible changes undergone by systems composed of a few particles – or indeed by composite many-particle systems moving coherently as in SQUIDS. If we pursue the view that we should learn our ways of thinking about natural phenomena from the way that nature behaves, then surely we should take very seriously the proposition that this division is a fundamental property of the physical world and not just a distinction we make for our own convenience. It is because of this that I find Prigogine's ideas so attractive. Instead of starting from the microscopic abstraction and trying to derive laws from it that will describe both microscopic and macroscopic phenomena, he suggests we do the opposite. Why not try taking as reality those processes in the physical world that are actually observed – the cat's death, the blackening photographic emulsion, the formation of a bubble in the liquid-hydrogen bubble chamber; and treat as 'illusion', or at least as an approximation to reality, the sub-atomic processes – the photon passing through both slits or the elementary particle changing by two quantum processes simultaneously? Of course, as I pointed out in Chapter 9, this implies a revolutionary change in our thinking: fundamental reality is not now the existence of the physical world, but the irreversible changes occurring in it – not 'being' but 'becoming'.

Reversible 'events' which by definition leave no permanent record are illusions – not just to us, but to the whole universe which undergoes no irreversible change as a result of their 'occurrence'. Alternatively, if we do attempt to make a realistic 'hidden-variable' description of such quantum processes, we should not be surprised that we can do so only if we accept non-locality. Space and time themselves are made manifest only through the irreversible processes so it is to be expected that if purely reversible 'happenings' have any reality this is outside this framework.

Another reason I find this way of looking at things attractive is the importance it gives to a traditional, serial way of considering the nature of time. Since Einstein it has become fashionable to look at time as just another dimension and to talk about 'space-time'. However, time and space are not equivalent concepts, even in the theory of relativity, and although we could imagine a world with a greater (or lesser) number of spatial dimensions than three, it is impossible to imagine a world with no time dimension. Without the possibility of change the idea of existence is meaningless, so for me at least there is no being without becoming. If, as a result of the modern work on irreversible processes, we were to be led to a fundamental physics that took as its central theme the idea that time really does flow in one direction, I at least should certainly welcome it.

I also like this approach to nature because it is fundamentally a scientific approach. Since the development of quantum theory, too many people, some of whom should have known better, have used it to open a door to some form of mysticism. The outrageous leap from the measurement problem to the necessity for the existence of the human soul is just one extreme example of this; another, as we saw at the end of Chapter 4, is the suggestion that the delocalization of quantum states can be used to 'explain' extra-sensory perception and all sorts of other 'paranormal' phenomena. Much of this is dispelled if the irreversible measuring processes are taken as the primary reality. Because, although often indeterministic, these processes are objectively real and are subject to locality and the rest of physics including the theory of relativity. Although we are a long way from any detailed understanding of the workings of the human brain, it is clear that the underlying processes are complex and chaotic and have more in common with measurement-type irreversible changes than with coherent pure quantum phenomena. It seems to me therefore that attempts to explain mental operations in traditional quantum terms are doomed to failure and that the idea sometimes put forward of a parallelism between quantum physics and psychology is only a superficial analogy at best. We are just beginning to understand the quantum behaviour of our chaotic universe; the hope should be that further study of this area will open up new possibilities of experimental test and that the distinction between what is illusion and what is reality can again be pursued by scientists as well as philosophers.

Further reading

This is not intended to be an exhaustive bibliography of the subject, but more a general guide to some of the larger number of relevant publications. A much more detailed survey of the literature is contained in the volume by Wheeler and Zurek cited below.

General

D. Bohm *Causality and Chance in Modern Physics* (Routledge and Kegan Paul, London, 1959 and 1984). This recently republished volume discusses the problems of quantum theory as they appeared thirty years ago to the leading proponent of hidden-variable theories.

P. C. W. Davies *Other Worlds* (Dent, London, 1980) discusses quantum ideas along with other developments in modern physics. Written by a professional physicist for the general reader.

B. d'Espagnat *Conceptual Foundations of Quantum Mechanics* (Benjamin, Massachusetts, 1976). This book and its author have played a large part in the revival of interest in quantum problems in recent years; written at a level suitable for the professional physicist competent at mathematics.

B. d'Espagnat *In Search of Reality* (Springer, New York, 1983). In this volume the author explains his recent ideas in a non-mathematical way.

A. I. M. Rae *Quantum Mechanics* (2nd ed. Hilger, Bristol, 1986). This university text book by the present author concentrates mainly on principles and applications, but also includes a chapter on the conceptual problems.

J. A. Wheeler and W. H. Zurek (eds) *Quantum Theory and Measurement* (Princetown University Press, 1983). These authors have collected together and reprinted a large number of the original articles published between 1926 and 1981. Of particular interest are the extracts from the Bohr–Einstein dialogue and the extensive bibliography.

Specific

Chapter 1:

J. Powers *Philosophy and the New Physics* (Meth
book provides a readable discussion of the deve
philosophical ideas over the last few centuries.

Chapter 2:

A. P. French and E. F. Taylor *An Introducti*
(Nelson, Middlesex, 1978). This university text
sion of the polarization properties of photons,
general, that goes a bit further than the present

Chapter 3:

J. F. Clauser and A. Shimony 'Bell's Theorem:
implications' (*Reports on Progress in Physics*,
1978). This review article contains a thorough
proofs of Bell's Theorem and describes the va
formed up to that time which was before the As

B. d'Espagnat 'The quantum theory and realit
Vol. 241, No. 11, pp. 128–66, 1979). This artic
account of Bell's Theorem and its implications fo
written before the Aspect experiments.

Chapter 5:

K. Popper and J. C. Eccles *The Self and its Brain* (
This is the book that sets out the ideas discusse
chapter.

D. R. Hofstadter and D. C. Dennett (eds) *T*
Middlesex, 1981). This collection of essays by a n
problem of consciousness contains a commentar
for a model of the mind based on the ideas of arti

J. Searle *Minds, Brains and Science* (BBC, 19
lectures argue against the separation of mind an
the idea of a computer-like model for the mind.

W. Sargeant and H. J. Eysenck *Explaining the Un*
and Nicholson, 1982). This survey of paranormal
chapter describing the attempts to explain them b

M. Gardner *Science: Good, Bad and Bogus* (Ox
1983). Although it contains few references to
collection of Martin Gardner's writings over the y
of robust criticisms of research into paranorma
refreshing antidote to the book by Sargeant and E

Further reading

This is not intended to be an exhaustive bibliography of the subject, but more a general guide to some of the larger number of relevant publications. A much more detailed survey of the literature is contained in the volume by Wheeler and Zurek cited below.

General

D. Bohm *Causality and Chance in Modern Physics* (Routledge and Kegan Paul, London, 1959 and 1984). This recently republished volume discusses the problems of quantum theory as they appeared thirty years ago to the leading proponent of hidden-variable theories.

P. C. W. Davies *Other Worlds* (Dent, London, 1980) discusses quantum ideas along with other developments in modern physics. Written by a professional physicist for the general reader.

B. d'Espagnat *Conceptual Foundations of Quantum Mechanics* (Benjamin, Massachusetts, 1976). This book and its author have played a large part in the revival of interest in quantum problems in recent years; written at a level suitable for the professional physicist competent at mathematics.

B. d'Espagnat *In Search of Reality* (Springer, New York, 1983). In this volume the author explains his recent ideas in a non-mathematical way.

A. I. M. Rae *Quantum Mechanics* (2nd ed. Hilger, Bristol, 1986). This university text book by the present author concentrates mainly on principles and applications, but also includes a chapter on the conceptual problems.

J. A. Wheeler and W. H. Zurek (eds) *Quantum Theory and Measurement* (Princetown University Press, 1983). These authors have collected together and reprinted a large number of the original articles published between 1926 and 1981. Of particular interest are the extracts from the Bohr–Einstein dialogue and the extensive bibliography.

Specific

Chapter 1:

J. Powers *Philosophy and the New Physics* (Methuen, London, 1982). This book provides a readable discussion of the development of scientific and philosophical ideas over the last few centuries.

Chapter 2:

A. P. French and E. F. Taylor *An Introduction to Quantum Physics* (Nelson, Middlesex, 1978). This university text book includes a discussion of the polarization properties of photons, and of polarized light in general, that goes a bit further than the present volume.

Chapter 3:

J. F. Clauser and A. Shimony 'Bell's Theorem: experimental tests and implications' (*Reports on Progress in Physics*, Vol. 41, pp. 1881–1927, 1978). This review article contains a thorough discussion of the various proofs of Bell's Theorem and describes the various experiments performed up to that time which was before the Aspect experiments.

B. d'Espagnat 'The quantum theory and reality' (*Scientific American*, Vol. 241, No. 11, pp. 128–66, 1979). This article gives a semi-popular account of Bell's Theorem and its implications for non-separability – also written before the Aspect experiments.

Chapter 5:

K. Popper and J. C. Eccles *The Self and its Brain* (Springer, Berlin, 1977). This is the book that sets out the ideas discussed in the first part of the chapter.

D. R. Hofstadter and D. C. Dennett (eds) *The Mind's I* (Penguin, Middlesex, 1981). This collection of essays by a number of writers on the problem of consciousness contains a commentary by the authors arguing for a model of the mind based on the ideas of artificial intelligence.

J. Searle *Minds, Brains and Science* (BBC, 1984). These 1984 Reith lectures argue against the separation of mind and brain but also oppose the idea of a computer-like model for the mind.

W. Sargeant and H. J. Eysenck *Explaining the Unexplained* (Weidenfeld and Nicholson, 1982). This survey of paranormal phenomena includes a chapter describing the attempts to explain them by quantum physics.

M. Gardner *Science: Good, Bad and Bogus* (Oxford University Press, 1983). Although it contains few references to quantum physics, this collection of Martin Gardner's writings over the years contains a number of robust criticisms of research into paranormal phenomena and is a refreshing antidote to the book by Sargeant and Eysenck cited above.

Chapter 6:

B. S. DeWitt and N. Graham (eds) *The Many-Worlds Interpretation of Quantum Mechanics* (Princeton University Press, 1973). This is a collection of a number of the original papers on this topic, including Everett's Ph.D. thesis. It is a pity that an extensive but less technical account of these ideas does not seem to be available.

Chapter 7:

A. J. Leggett 'Schrödinger's cat and her laboratory cousins' (*Contemporary Physics*, Vol. 25, pp. 583–98, 1984). This recent review article describes Professor Leggett's idea of a possible macroscopic resolution of the measurement problem and how it may be tested by experiments on SQUIDS.

Chapter 8:

P. C. W. Davies *Space and Time in the Modern Universe* (Cambridge, 1977). This wide-ranging semi-popular account of modern physics includes an excellent discussion of irreversibility and time asymmetry.

P. T. Landsberg (ed.) *The Enigma of Time* (Hilger, Bristol, 1982). This collection of papers spans many aspects of the nature of time, including irreversibility.

Chapter 9:

I. Prigogine *From Being to Becoming* (Freeman, San Francisco, 1980). This book forms much of the basis of the discussion in the chapter as well as describing a number of other interesting manifestations of irreversible thermodynamics.

I. Prigogine and I. Stengers *Order out of Chaos* (Heinemann, 1984). A recent popular version of the book cited above.

Index